Reliability Prediction and Testing Textbook

Reliability Prediction and Testing Textbook

Lev M. Klyatis
Professor Emeritus
Habilitated Dr.-Ing., Dr. of Technical Sciences, PhD

Edward L. Anderson
BS in Mechanical Engineering

Registered Office
John Wiley & Sons, Inc., 111 River Street, Hoboken, NJ 07030, USA

Editorial Office
111 River Street, Hoboken, NJ 07030, USA

For details of our global editorial offices, customer services, and more information about Wiley products visit us at www.wiley.com.

Library of Congress Cataloging-in-Publication Data

Names: Klyatis, Lev M., author. | Anderson, Edward L., 1945- editor.
Title: Reliability prediction and testing textbook / by Lev M. Klyatis ;
   Edward L. Anderson, language editor.
Description: Hoboken, NJ, USA : Wiley, 2018. | Includes bibliographical
   references and index. |
Identifiers: LCCN 2017054872 (print) | LCCN 2017059378 (ebook) | ISBN
   9781119411925 (pdf) | ISBN 9781119411932 (epub) | ISBN 9781119411888
   (cloth)
Subjects: LCSH: Accelerated life testing.
Classification: LCC TA169.3 (ebook) | LCC TA169.3 .K5964 2018 (print) | DDC
   620/.00452–dc23
LC record available at https://lccn.loc.gov/2017054872

Cover Design: Wiley
Cover Image: © naqiewei/Getty Images

Set in 10/12pt Warnock by SPi Global, Chennai, India

Printed in the United States of America

VWEP44369_111418

DEDICATION

To my wife Nellya Klyatis

To my wife Carol Anderson

Contents

Preface

Lev M. Klyatis and Edward L. Anderson

When Lev Klyatis began his engineering career in 1958 as a test engineer at the Ukrainian State Test Center for farm machinery, he was surprised to learn that, even after extensive testing by this center, the testing was not accurately predicting the reliability of the products as used by farmers. This test center would conduct farm machinery field testing during one season of operation, and make the recommendation to manufacture the new product based on results of this single-season testing.

Neither the designers, nor test engineers, nor the researchers, nor other decision-makers involved knew what would happen after the first season. The test center was not accurately predicting true product reliability during the life cycle of the machines. Later, Lev Klyatis realized that this situation was not unique to farm machinery, but was related to other areas of industry and other countries over the world, even when they claimed to be doing accelerated reliability testing.

Why are we writing this book? As will be seen, it is the author's observation that the developments of technology, methodologies, hardware, and software are advancing at an unprecedented rate. But, in the same time, we find that reliability testing and prediction are advancing much more slowly; and in many cases it is common to find reliability testing and prediction methodologies that have changed little in the past 60–70 years. As product complexity increases, the need for near-perfect product reliability, which is founded on the ability to accurately predict reliability prior to widespread production and marketing, becomes a company's critical objective. Failure to predict and remedy failures can result in human tragedy, as well as serious financial losses to the company. Consider the two following recent examples.

On May 31, 2009, Air France's flight AF447 departed Rio de Janeiro en route to Paris carrying 228 passengers and crew; several hours into the flight it crashed into the Atlantic Ocean, killing all on board [1, 2]. A contributing factor in the accident was pitot tubes, which were believed to have iced, resulting in the loss of accurate airspeed and altitude information. The pitot tubes were known to have a problem with icing and had been replaced by several other airlines. Following the accident, The European Aviation Safety Agency (EASA) made compulsory the replacement of two out of three airspeed pitot's on Airbus A330s and A340s AD (204-03-33 Airbus Amendment 3913-447. Docked 2001- NM-302-AD), and the FAA followed with a near-identical requirement in promulgating Docket No. FAA-2009-0781 AD 2009-18-08 Final Rule Airworthiness Directive AD concerning Airbus A330 and A340 airplanes. It is profoundly troubling that in age of state-of-the-art fly-by-wire jet aircraft, we would be encountering problems with pitot tube icing [3].

In February 2014, General Motors issued a recall for over 2.6 million vehicles to correct an ignition switch defect responsible for at least 13 deaths, and possible more than 100, and this does not include those seriously injured. The ignition switch could move from the "On" position to the "Acc" position; and, when this happened, safety systems, such as air bags, anti-lock brakes, and power steering, could be disabled with the vehicle moving. The problem was initially uncovered by GM as early as 2001, with continued recommendations to change the design through 2005, but this recommendation was rejected by management.

By the end of March of 2015 the cost to GM for the ignition switch recalls was $200 million and was expected to reach as much as $600 million [4–6]. Add to the financial loss the personal tragedy of those killed or injured and to their families, and the true cost of failed reliability prediction becomes evident. By the end of the next decade it is almost a certainty that you will be sharing the road with some type of autonomous vehicle [7–9]. Consider the degree of reliability prediction that will be needed to provide the level of confidence needed. Whether you are driving an autonomous vehicle or merely sharing the road with them, you are literally betting your life on the adequacy and accuracy of the reliability testing for each critical component and decision-making process. Considering that, today, we are having difficulties with ignition switches and pitot tubes, this will be a major undertaking.

This is particularly so when the testing will need to account for such varied environmental conditions as heat, cold, rain, snow, roadway salt, and various other expected and unexpected contaminants. Couple this with the 10 years plus life of the average automobile [10], and reliability assurance against a wide variety of degradations is necessary, and all life failure modes must default to a fail-safe mode. These are only examples from many real-life problems that are connected with inadequate reliability prediction and testing methods.

Unfortunately, too often these costs for failed reliability prediction and testing are never factored into an organization's decision-making processes. While the human and financial impacts of responsible new product development should be foremost in an organization's (including research and pre-design, and testing) activities and concerns, too often they are overlooked or assumed to be someone else's responsibility, especially in a large organization. But if we are to remain a civilized society, such responsibility cannot and should not be delegated up the chain of command.

One of the major concerns is the development of higher speed and processing power of electronic developments occurring at a much greater rate than other areas of people's activity. In fact, Moore's law—the expectation that microprocessor power doubles every 18 months—is widely accepted in the industry. How often do we consider the effects and the implications of electronic developments, especially in new software development, that transfers thinking resulting in real brain development to what may be termed virtual thinking? Virtual thinking is surrendering human thinking and mental development to a system of control that provides answers automatically, with a minimal role for thought. Consider how calculators and electronic cash registers have reduced people's ability to do basic mathematics; or how GPS navigation has diminished the average person's ability to read a map and plot the course to their destination—it is so much easier to just type in the destination address and let the machine direct you turn by turn to the destination. But the development of thinking skills and using them for the advancement of society and civilization are the basic differences between humans and animals.

Unfortunately, people often do not understand that electronic systems are only part of a system of controls containing real physical limits, processes and technologies, and that there are limits to what can be accomplished with software and corresponding hardware. The most advanced automotive stability system cannot allow the vehicle to corner at an unsafe speed. While the system may enhances a person's abilities as a driver, it cannot violate the rules of physics. And frequently, the enhancement provided by technology is accompanied by a reduction in the skill of the operator as they become increasingly dependent on the technology.

A common example of this occurs when predictions are based on abstract (virtual) processes which are different than the real (actual) processes. Too often, prediction reliability is based on only virtual (theoretical) understanding and does not account for the real situations in people's real life. Because of this, many reliability prediction approaches are based only on theoretical knowledge, and the testing used is not a real interconnected process, but relies on secondary conditions expected in their virtual world. Therefore, testing development, together with prediction development, is not developing as quickly as needed and is moving forward very slowly, much more slowly than design and manufacturing processes (Figure 3.6). Reliability (mostly accelerated reliability) testing needs technology, equipment, and corresponding costs as complicated as the new products they are testing. But too often should be key concerns the management of many companies prefer to pay as little as they can for this technology and the development of necessary testing equipment. They want to save expenses for this stage of product development. Phillip Coyle, the former director of the Operational Test and Evaluation Office (Pentagon) said in the US Senate that if, during the design and manufacturing of complicated apparatus such as a satellite, one tries to save a few pennies in testing, the end results may be a huge loss of thousands of dollars due to faulty products which have to be replaced because of this mistake. This relates to other products, too.

As a result, product reliability prediction is unsuccessful, which is reflected in increasing and unpredictable recalls, decreasing actual reliability of industrial products and technologies, decreasing profit and increasing life-cycle cost, in comparison with that planned during design. Lev Klyatis came to realize that the reliability prediction approaches utilized at the time (and even frequently now) did not obtain accurate or adequate initial information to successfully predict the reliability of the machinery in actual use. There was a need for more extensive and accurate testing prior to manufacturing if long-term field reliability was a desired outcome.

Testing needed to account for multiple seasons, long life, varying operational conditions, and other factors if the machinery was to perform satisfactorily. This meant a refocusing of the accelerated testing (laboratory, proving grounds, field intensive use), initially for farm machinery and then for other products and technological systems. Without accurate simulation of real- world conditions, test results may be very different from real-world results. If one thinks that the approach of simply recalculating proving ground (or any other simple simulation of real life) results is all that is needed to provide real world simulation, they will be disappointed. If the initial information used in testing (testing protocols) is not correct, the prediction method will not be useful in practice. But, it also became apparent that reliability prediction approaches were mostly theoretical in nature, because of the difficulty of accurately simulating the real-world situation.

Much of this continues even now. As a result, companies design, begin manufacturing, and have the product in the hands of the customers only to discover reliability issues that result in serious consequences—usually economic losses, but increasingly legal consequences. And, all too often, it is the consumer who suffers from the poor reliability prediction. As an example, Figure 1 shows automotive recalls for the years between 1980 and 2013. While automobile technology is relatively mature, product complexity and technology changes are resulting in significant amounts of recalls, which is indicative of testing adequacy moving too slowly (see Figure 3.6).

This situation continues. For example [12]:

> In September (2016), Ford expanded its recall on hundreds of thousands of Ford and Lincoln vehicles early in the year to a staggering 2,383,292 vehicles. The problem involves a side door-latching component, which results in the door not closing or latching properly. The door will either not close at all, or the door could close temporarily, only to reopen later

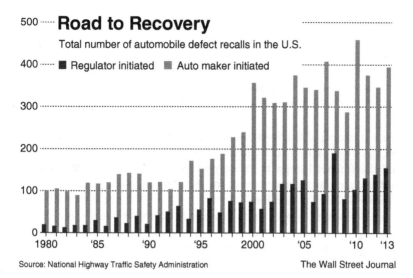

Figure 1 Total number of automotive recalls in the USA in 1980–2013 [11] (vertical line is percent, if the number of recalls is equivalent 100% in 1980, in 2010 number of recalls in percentage was approximately 500%). *Source:* National Highway Traffic Safety Administration.

while the vehicle is in motion. This, of course, would be a terrifying thing to encounter and severely increases the risk of injury. Needless to say, Ford and Lincoln dealers will replace the side door latches at no cost to the consumer.

But the question remains: Why should we be having problems with something as basic as a door latching system? And [13]:

Since the beginning of the year, Toyota has recalled nearly 3 million RAV4s and RAV4 EVs after evidence emerged that there was an issue with the second-row safety belts. The problem involves how the seat belt interacts with the metal frame of the seat cushion, and, in the event of an accident, the metal frame could cut the seat belt right in two. As a result the seat belt fails at its most fundamental function. In order to alleviate this issue, Toyota simply adds a cover to the metal seat cushion frame that prevents it from cutting through the belt in a collision.

And again, the question is posed: Why was this not discovered in testing? In 2017, Reuters informed [14]:

The U.S. Transportation Department said automakers recalled a record high 53.2 million vehicles in 2016 in the United States in part because of a massive expansion in call back to replace Takata Corp 7312.T air bag inflators.

Under aggressive enforcement ..., automakers issued a record-setting 927 recall campaigns, up 7 percent over the previous high set in 2015. Last year's recall of 53.2 million total vehicles topped the previous all-time high of 51.1 million set in 2015, the department said.

While there are many publications discussing automotive and other industrial product recalls, most of them focus on the reliability and safety aspects of these recalls, and the financial impacts to the parties involved. But reliability and safety problems are not the true causes for these recalls. It is most often the result of faulty reliability prediction.

The low level of reliability prediction can lead to deaths or injuries as a result of incidents, and ultimately result in increasing cost of the product, decreasing the company's profit and image, incurring losses to the customers, and many other problems. In some cases, low-level reliability prediction can even result in criminal prosecution of the company's leaders or staff. These problems are considered in greater detail in the book *Successful Prediction of Product Performance* [11]. Finally, one should not forget that poor product reliability is connected to other performance factors, such as durability, maintainability, safety, life-cycle cost, profit, and others. In the real world, these are often interconnected and interact with each other.

It is the purpose of this textbook to provide guidance on how one can improve reliability prediction and testing as two interconnected components with corresponding close development for both.

Lev M. Klyatis
Edward L. Anderson

References

1 AirSafe.com, LLC. (2011). *Air France A330 crash in the Atlantic Ocean.* http://www.airsafe.com/plane-crash/air-france-flight-447-airbus-a330-atlantic-ocean.htm (accessed February 5, 2018).

2 Traufett G. (2010). *The last four minutes of Air France Flight 447.* http://www.spiegel.de/international/world/death-in-the-atlantic-the-last-four-minutes-of-air-france-flight-447-a-679980.html (accessed February 5, 2018).

3 The Associated Press. (2009). *Airspeed sensor troubled history ignored?* https://www.cbsnews.com/news/airspeed-sensor-troubled-history-ignored/ (accessed February 5, 2018).

4 Shepardson D, Burden M. (2015). *GM to pay $900M to resolve ignition switch probe.* http://www.detroitnews.com/story/business/autos/general-

motors/2015/09/16/gm-billion-fine-resolve-ignition-switch-probe/32526023/ (accessed February 5, 2018).

5 Valcies-Dapena P, Yellin T. (2017). *GM: Steps to a recall nightmare.* http://money.cnn.com/infographic/pf/autos/gm-recall-timeline/index.html (accessed February 5, 2018).

6 NastLaw, LLC (n.d.). *General Motors recall timeline.* https://nastlaw.com/general-motors-recall/timeline/ (accessed February 5, 2018).

7 BI Intelligence. (2016). *10 million self-driving cars will be on the road by 2020.* http://uk.businessinsider.com/report-10-million-self-driving-cars-will-be-on-the-road-by-2020-2015-5-6 (accessed February 5, 2018).

8 Insurance Journal. (2015). *How the age of autonomous vehicles will evolve.* http://www.insurancejournal.com/news/national/2015/04/24/365573.htm (accessed February 5, 2018).

9 Muller J. (2015). *The road to self-driving cars: a timeline.* https://www.forbes.com/sites/joannmuller/2015/10/15/the-road-to-self-driving-cars-a-timeline (accessed February 5, 2018).

10 Rechtin M. (2013). *Average age of U.S. car, light truck on road hits record 11.4 years, Polk says.* http://www.autonews.com/article/20130806/RETAIL/130809922/average-age-of-u.s.-car-light-truck-on-road-hits-record-11.4-years (accessed February 5, 2018).

11 Klyatis L. (2016). *Successful Prediction of Product Performance. Quality, Reliability, Durability, Safety, Maintainability, Lifecycle Cost, Profit, and Other Components.* SAE International.

12 Boudette NE, Kachi H. (2014). *Big car makers in race to recall.* https://www.wsj.com/articles/toyota-recalls-6-4-million-vehicles-worldwide-1397025402 (accessed February 5, 2018).

13 Harrington J. (2016). *Total recall: 5 of the most expansive automotive recalls of 2016.* https://blog.cargurus.com/2016/11/04/5-of-the-most-expansive-automotive-recalls-of-2016 (accessed February 5, 2018).

14 Shepardson D, Paul F. (2017). *U.S. auto recalls hit record high 53.2 million in 2016.* https://uk.reuters.com/article/us-usa-autos-recall/u-s-auto-recalls-hit-record-high-53-2-million-in-2016-idUKKBN16H27A (accessed February 5, 2018).

About the Authors
Lev M. Klyatis

Lev Klyatis is a senior adviser at SoHaR, Incorporated. He holds three doctoral degrees: engineering technology, PhD; engineering technology, ScD (a high-level East European doctoral degree); and engineering, Habilitated Dr.-Ing. (a high-level west European doctoral degree).

His major scientific/technical expertise has been in the development of new directions for the successful prediction of product performance, including reliability during service life (or other specified time), and accelerated reliability and durability testing with accurate physical simulation of field conditions. Dr. Klyatis developed new ideas and unique approaches to accurate simulation. This approach was founded on the integration of field inputs, safety, and human factors, improvement in the engineering culture, and accelerated product development. He developed a methodology for reducing complaints and recalls. His approach has been verified in various industries, primarily automotive, farm machinery, airspace, and aircraft. He has served as a consultant to Ford, DaimlerChrysler, Nissan, Toyota, Jatco Ltd, Thermo King, Black and Decker, NASA Research Centers, Carl Schenk (Germany), and many others.

He was qualified as a full professor by the USSR's Highest Examination Board and was a full professor at the Moscow University of Agricultural Engineers. He has served on the US–USSR Trade and Economic Council, the United

Nations Economic Commission for Europe, and the International Electrotechnical Commission (IEC). He also served as an expert of the United States, and an expert of the International Standardization Organization and International Electrotechnical Commission (ISO/IEC) Joint Study Group in Safety Aspects of Risk Assessment. He was the research leader and chairman of the State Enterprise Testmash, Moscow, Russia, and principal engineer of a state test center. He is presently a member of the World Quality Council, the Elmer A. Sperry Board of Award, SAE International G-11 Reliability Committee, the Integrated Design and Manufacturing Committee of SAE International World Congresses, Session Chairman for SAE World Congresses in Detroit since 2012, and a member of the Governing Board of SAE Metropolitan Section. He has been a seminar instructor for the American Society for Quality.

Lev Klyatis is the author of over 250 publications, including 12 books, and holds more than 30 patents worldwide. Dr. Klyatis frequently speaks at technical and scientific events that are held around the world.

Edward L. Anderson

Edward Anderson is a professional engineer with over 40 years' experience in the design, procurement, and operation of highly specialized automotive vehicles. He is active in the engineering profession in SAE International, the Sperry Board of Award, and the SAE G-15 Airport Ice and Snow Control Equipment Committee, of which he was a founding member. Ed studied engineering at Newark College of Engineering, graduating with a BS in mechanical engineering and a commission in the US Air Force. He served as an Air Force pilot for 5 years, logging over 1000 hours flying time in C-141

Starlifter four-engine turbojet transports flying global airlift missions, and several hundred hours in HH43 Huskie rescue helicopters.

After the Air Force, Ed worked in the private sector as an engineer in the design and manufacture of fire apparatus, small buses, tank truck vehicles, and assorted custom mobile equipment, and as transportation engineer for GPU Service, an electric utility. In 1980, he joined the Port Authority of NY & NJ, starting as an automotive engineer and progressing to supervisor of automotive engineers. The Port Authority's fleet of approximately 3000 vehicles is extremely diverse, including everything from large airport emergency response, security, and snow removal equipment to very compact emergency vehicles that must operate within the confines of the agency's bridges and tunnels. Ed's engineering group was fully involved in all aspects of acquiring suitable vehicles and equipment to support these operations, and was also responsible for boats, emergency generators, and fire pump procurements for the Port Authority's diverse facilities. His group is also responsible for technical analysis, failure investigations, and recommendations for future applications in these areas throughout the agency.

Ed has done presentations to SAE International, the International Aviation Snow Symposium, NAFA, and other groups on the art and science of procuring complex, highly custom mission-critical vehicles and equipment. As an aid to training new engineers in his group, as well as related management staff, he developed and teaches the course "Engineering the Motor Truck" to teach the complexities and key decision points in this rather specialized field of engineering. The course covers fundamental differences in the specifying of a vocational truck, key analytical factors such as legal limitations, weight and balance analysis, power train considerations, and factors specific to the vocational application of the truck. In addition to teaching his staff, Ed has served as Chair for the SAE Metropolitan Section.

Ed is a professional engineer with a master's degree in health and safety engineering from the New Jersey Institute of Technology, and a post master's certificate from Dowling College in total quality management.

Introduction
Lev M. Klyatis

What is reliability?

Reliability as used in this text is the science aimed at predicting, analysing, preventing, and mitigating failures over time. Reliability is quality over time. A reliable, trouble-free product continues to satisfy customers for a long time. In the narrow sense, reliability is the probability that a device will operate successfully for a specific period of time and under specified conditions when used in a manner consistent with the purpose intended.

It is the authors' contention that the money invested in adequate testing during design and production for developing a reliable product will result in greater profitability during both the warranty period and the service life of the product.

Who gains from improved reliability?

The short answer is everyone, specifically:

- **The manufacturer**, because their product has additional customer appeal, through higher quality, improved in-service time, faster service, lower cost of recalls, and a reduced need for changes in the manufacturing processes.
- **The customer**, because the equipment purchased performs as expected, is easier and less costly to support, and has reduced out-of-service time (better availability).
- **Society**, because there are reduced deaths, injuries, lost productivity, and associated impact resulting directly or indirectly from failures of the product.

Reliability-related costs are not just the costs of reliability testing, but include all the costs resulting from product field failures or perceived failures from the time of shipment throughout the life of the product. They include warranty costs, recall costs, design change cost, manufacturing change cost, and the opportunity cost of lost customers.

Improvements in reliability made early in the equipment life cycle may also result in total life-cycle costs saving (significant decreasing of maintenance and inventory costs).

Failure analysis is the process of collecting and analysing data to determine the cause of a failure and developing methods of remediation. Failure analysis is an important aspect of many branches of manufacturing, especially in electronics, where it is a vital tool used in the development of new products and for the improvement of existing products. Effective failure analysis is especially important in the manufacturing and use of life-safety and mission-critical equipment. Of course, it is also important for other areas of industry. Failure analysis may be applied to both products and processes. Failure analysis may be conducted at the design stage, manufacturing, or field-use stage of the product life cycle.

While successful prediction is a key element in product development, too often reliability prediction has been demonstrated to be less than successful. Historically, the term "prediction" has been used to denote the process of applying mathematical models and data to estimate the field performance of a product or process. This was frequently done before empirical data were available for the product or system. Prediction is successful if it prevents unexpected failures of the product or improves the product's reliability characteristics.

1

Analysis of Current Practices in Reliability Prediction

Lev M. Klyatis

1.1 Overview of Current Situation in Methodological Aspects of Reliability Prediction

Because the problem of reliability prediction is so important there are many publications in the area of the methods of reliability prediction, mostly in the area of electronics. This is especially important when the reliability prediction is necessary to provide a high degree of effectiveness for the products.

Most of these traditional methods of reliability prediction utilize failure data collected from the field to estimate the parameters of the failure time distribution or degradation process. Using this data one can then estimate the reliability measures of interest, such as the expected time to failure or quantity/quality of degradation during a specified time interval and the mean time to failure (or degradation).

Reliability prediction models that utilize accelerated degradation data, rather than those models that utilize degradation data obtained under normal conditions, are often performed because the degradation process is very slow under normal field conditions and prediction modeling based on normal conditions would require excessive time.

One example of such practices is the procedure used by Bellcore for both hardware and software. O'Connor and Kleyner [1] provide a broad description of a reliability prediction method that emphasizes reliability prediction consisting of outputs from the reliability prediction procedure:

- steady-state failure rate;
- first year multiplier;
- failure rate curve;
- software failure rate.

The output from the reliability prediction procedure is various estimates of how often failure occurs. With hardware, the estimates of interest are the steady-state failure rate, the first year multiplier, and the failure rate curve. The steady-state failure rate is a measure of how often units fail after they are more than 1 year old. The steady-state failure rate is measured in FIT (failures per 10^9 h of operation). A failure rate of 5000 FIT means that about 4% of the units that are over a year old will fail during following one year periods. The first year multiplier is the ratio of the failure rate in the first year to that in subsequent years. Using these, one generates a failure rate curve that provides the failure rate as a function of the age (volume of work) of the equipment. For software, one needs to know how often the system fails in the field as a result of faults in the software.

The Bellcore hardware reliability prediction is primarily designed for electronic equipment. It provides predictions at the device level, the unit level, and for simple serial systems, but it is primarily aimed at units, where units are considered nonrepairable assemblies or plug-ins. The goal is to provide the user with information on how often units will fail and will need to be replaced.

The software prediction procedure estimates software failure intensity and applies to systems and modules.

There are many uses for reliability prediction information. One such example is, it can be used as inputs to life-cycle cost or profit studies. Life-cycle cost studies determine the cost of a product over its entire life. Required data include how often a unit will have to be replaced. Inputs to this process include the steady-state failure rate and the first year multiplier.

The Bellcore reliability prediction procedure consists of three methods [1]:

1. *Parts count.* These predictions are based solely by adding together the failure rates for all the devices. This is the most commonly used method, because laboratory and field information that is needed for the other methods is usually not available.
2. *Incorporating laboratory information.* Device or unit level predictions are obtained by combining data from a laboratory test with the data from the parts count method. This allows suppliers to use their data to produce predictions of failure rates, and it is particularly suited for new devices for which little field data are available.
3. *Incorporating field information.* This method allows suppliers to combine field performance data with data from the parts count method to obtain reliability predictions.

Mechanical reliability prediction [2] uses various stress factors under operating conditions as a key to reliability prediction for different devices. This situation is more common in testing mechanical devices, such as bearings, compressors, pumps, and so on than with electronic hardware.

Although, the number of factors that appear to be needed in reliability testing calculations may appear excessive, tailoring methods can be used to remove factors that have little or no influence, or for which limited data are available. Generally, the problems encountered in attempting to predict the reliability of mechanical systems are the lack of:

- specific or generic failure rate data;
- information on the predominant failure modes;
- information on the factors influencing the reliability of the mechanical components.

The mechanical reliability prediction approach can be useful if there is a close connection with the source of information for calculation reliability during the desired time, whether that be warranty period, service life, or other defined period. Obtaining accurate initial information is a critical factor in prediction testing.

Unfortunately, the approaches described earlier, and other prediction methods, provide little guidance on how one can obtain accurate initial information that simulates real product reliability over time or amount of use (volume of work, or duty cycle). Without accurate simulation information, its usefulness is minimal.

Proper understanding of the role of testing and the requirement to do this testing before the production and use of a product is critical and can easily lead to poor product reliability prediction that will negatively impact financial performance.

Prediction is only useful if it reduces early product degradation and prevents premature failures of the product.

There are many recent publications addressing electronics, automotive, and other product recalls. While they usually address reliability failures in terms of safety matters that affect peoples lives by contributing to deaths or injuries, they may also consider economic impacts.

As was mentioned previously, such reliability and other problems are results, not causes. The actual causes of these recalls, and many other technical and economic problems, were a direct result of the inefficient or inadequate prediction of product reliability during the design, and prior to, the manufacturing process.

In the end, it is poorly executed prediction that negatively impacts the organizations financial performance.

Therefore, while many popular and commonly used approaches appear to be theoretically interesting, in the end they do not successfully predict reliability for the product in real-world applications.

Consider the consequences of the recalls of Takata automobile air bag inflators [3]:

> So far, about 12.5 million suspect Takata inflators have been fixed of the roughly 65 million inflators (in 42 million vehicles) that will ultimately be affected by this recall, which spans 19 automakers. Carmakers and federal officials organizing the response to this huge recall insist that the supply chain is churning out replacement parts, most of which are coming from companies other than Takata. For those who are waiting,

NHTSA advises that people not disable the airbags; the exceptions are the 2001–2003 Honda and Acura models that we listed on this page on June 30, 2016—vehicles which NHTSA is telling people to drive only to a dealer to get fixed.

Meanwhile, a settlement stemming from a federal probe into criminal wrongdoing by Takata is expected early next year—perhaps as soon as January—and could approach $1 billion.

A key to preventing these situations is the use of advanced test methods and equipment; that is, accelerated reliability testing (ART) and accelerated durability testing (ADT). Implementation of these systematic procedures greatly helps in assuring successful prediction of industrial product reliability.

It is also true that advances in technology generally result in more complicated products and increased economic development costs. Such advances require even more attention to accurately predict product reliability.

When performed successfully, prediction is beneficial to all stages of the product life cycle: start-up, production and manufacturing, warranty, and long-term aftermarket support. It touches the lives of all concerned (designers, suppliers, manufacturers, customers), and often even uninvolved third parties who may be affected by the product's failure. It also provides the mechanism for product improvement at any time from the earliest stages of R&D throughout the entire product life cycle.

Currently there are many other publications mostly related to the theoretical aspects of reliability prediction. Many of primarily relate to failure analysis. Some popular failure analysis methods and tools include:

- Failure Reporting, Analysis, and Corrective Action System (FRACAS);
- Failure Mode, Effects and Critical Analysis (FMECA);
- Failure Mode and Effects Analysis (FMEA);
- Fault Tree Analysis (FTA).

FavoWeb is Advanced Logistic Developments' (ADL's) third-generation, web-based and user-configurable FRACAS that captures information about equipment or the processes throughout its life cycle, from design, through to production, testing, and customer support.

FavoWeb has been adopted by world-class organizations who, for the first time ever, implement a FRACAS application that seamlessly communicates with any given enterprise resource planning (ERP) system (SAP, ORACLE, MFGpro, etc.), while proving a user-friendly and flexible, yet robust, failure management, analysis, and corrective action platform.

The FavoWeb FRACAS features include:

- full web-base application;
- user permission mechanism—complies with International Traffic in Arms Regulations requirements;

- flexible, user configurable application;
- seamless communication with ERP/product data management/Excel/Access and other legacy systems;
- web services infrastructure;
- failure/event chaining and routing;
- compatible with PDAs;
- voice-enabled failure reporting;
- advanced query engine for user-defined reports.

It also allows the user to decompose the system or process into components or subprocesses. And, for each functional block, it allows the user to define name and function, and enter failure mode causes and effects manually or from libraries. The "Process & Design FMEA" module provides full graphical and textual visibility of the potential failure mode → cause → effects chain.

1.1.1 What is a Potential Failure Mode?

Potential failure mode is any manner in which a component, subsystem, or system could potentially fail to meet the design intent. The potential failure mode could also be the cause of a potential failure mode in a higher level subsystem or system, or be the effect of a potential failure.

Reliability in statistics and psychometrics is the overall consistency of a measure. A measure is said to have a high reliability if it produces similar results under consistent conditions [4]:

> It is the characteristic of a set of test scores that relates to the amount of random error from the measurement process that might be embedded in the scores. Scores that are highly reliable are accurate, reproducible, and consistent from one testing occasion to another. That is, if the testing process were repeated with a group of test takers, essentially the same results would be obtained. Various kinds of reliability coefficients, with values ranging between 0.00 (much error) and 1.00 (no error), are usually used to indicate the amount of error in the scores.

For example, measurements of people's height and weight are often extremely reliable [4, 5].

There are several general classes of reliability estimates:

- *Inter-rater reliability* assesses the degree of agreement between two or more raters in their appraisals.
- *Test–retest reliability* assesses the degree to which test scores are consistent from one test administration to the next. Measurements are gathered from a single rater who uses the same methods or instruments and the same testing conditions [6]. This includes intra-rater reliability.

- *Inter-method reliability* assesses the degree to which test scores are consistent when there is a variation in the methods or instruments used. This allows inter-rater reliability to be ruled out. When dealing with forms, it may be termed *parallel-forms reliability* [6].
- *Internal consistency reliability*, assesses the consistency of results across items within a test [6].

A test that is not perfectly reliable cannot be perfectly valid, either as a means of measuring attributes of a person or as a means of predicting scores on a criterion. While a reliable test may provide useful valid information, a test that is not reliable cannot be valid [7].

1.1.2 General Model

Recognizing that, in practice, testing measures are never perfectly consistent, theories of statistical test reliability have been developed to estimate the effects of inconsistency on the accuracy of measurement. The starting point for almost all theories of test reliability is the concept that test scores reflect the influence of two types of factors [7]:

1. *Factors that contribute to consistency*—stable characteristics of the individual or the attribute that one is trying to measure.
2. *Factors that contribute to inconsistency*—features of the individual or the situation that can affect test scores but have nothing to do with the attribute being measured.

These factors include [7]:

- Temporary, but general, characteristics of the individual—health, fatigue, motivation, emotional strain;
- Temporary and specific characteristics of the individual—comprehension of the specific test task, specific tricks or techniques of dealing with the particular test materials, fluctuations of memory, attention or accuracy;
- Aspects of the testing situation—freedom from distractions, clarity of instructions, interaction of personality, sex, or race of examiner;
- Chance factors—luck in selection of answers by sheer guessing, momentary distractions.

1.1.3 Classical Test Theory

The goal of reliability theory is to estimate errors in measurement and to suggest ways of improving tests so that these errors are minimized. The central assumption of reliability theory is that measurement errors are essentially random. This does not mean that errors arise from random processes. For any individual, an error in measurement is not a completely random event. However, across a large number of individuals, the causes of measurement error are assumed to be so varied that measure errors act as random variables. If errors have the essential characteristics of random variables, then it is reasonable to assume that errors are equally likely to be positive or negative, and that they are not correlated with true scores or with errors on other tests.

$$\rho_{xx'} = \frac{\sigma_T^2}{\sigma_X^2} = 1 - \frac{\sigma_E^2}{\sigma_X^2}$$

Unfortunately, there is no way to directly observe or calculate the *true score*, so a variety of methods are used to estimate the reliability of a test. Some examples of the methods used to estimate reliability include test–retest reliability, internal consistency reliability, and *parallel-test reliability*. Each method approaches the problem of accounting for the source of error in the test somewhat differently.

1.1.4 Estimation

The goal of estimating reliability is to determine how much of the variability in test scores is due to errors in measurement and how much is due to variability in true scores. There are several strategies, including [7]:

1. *Test–retest reliability method.* Directly assesses the degree to which test scores are consistent from one test administration to the next. It involves:
   - administering a test to a group of individuals;
   - readministering the same test to the same group at some later time; and
   - correlating the first set of scores with the second.
   The correlation between scores on the first test and the scores on the retest is used to estimate the reliability of the test using the Pearson product-moment correlation coefficient; see also *item-total correlation* in Ref. [7].
2. *Parallel-forms method.* The key to this method is the development of alternate test forms that are equivalent in terms of content, response processes and statistical characteristics. For example, alternate forms exist for several tests of general intelligence, and these tests are generally seen as equivalent [7].

With the parallel test model it is possible to develop two forms of a test that are equivalent, in the sense that a person's true score on form A would be identical to their true score on form B. If both forms of the test were administered to a number of people, differences between scores on form A and form B may be due to errors in measurement only [7].

This method treats the two halves of a measure as alternate forms. It provides a simple solution to the problem that the *parallel-forms method* faces: the difficulty in developing alternate forms [7]. It involves:

- administering a test to a group of individuals;
- splitting the test in half;
- correlating scores on one half of the test with scores on the other half of the test.

There are many situations in which one needs to make a prediction about a product's performance before the product is in production. This means prediction is needed prior to production or warranty data being available for analysis. Many companies have product development programs that require design engineers to produce designs that will meet a certain reliability goal before the project is permitted to move on to the following phases (building prototypes, pre-manufacturing, and full manufacturing). This is to avoid committing the business to investing significant resources to a product with unproven reliability before leaving the design stage. This is especially difficult because a new design can involve components or subsystems that have no previous testing, and have no history of being used in the field by customers. Often, they encompass totally new items and not redesigned components or subsystems of existing components which would have prior histories [7].

In other cases, companies that may not have the capabilities, resources, or time to test certain (noncrucial) components/subsystems of a system, but still need to use some estimates of the failure rate of those components to complete their system reliability analysis.

Lastly, manufacturers are often required to submit reliability predictions usually based on a specific prediction standard with their bid or proposal for a project.

As was written in Ref. [7], the following are a few advantages and disadvantages related to standards based reliability prediction.

The advantages of using standards-based reliability prediction are:

- They can help to complete the system reliability block diagrams (RBDs) or FTAs when data for certain components/subsystems within the system are not available.
- It is sometimes accepted and/or required by government and/or industry contracts for bidding purposes.

The disadvantages of using standards-based reliability prediction are:

- Reliance on standards that may not reflect the products actual performance.
- Although standards-based reliability prediction addresses prediction under different usage levels and environmental conditions, these conditions may not accurately reflect the products actual application.
- Some of the standards are old and have not been updated to reflect the latest advances in technologies.
- The result from such predictions is a constant failure rate estimation that can only be used within the context of an exponential reliability model

(i.e., no wearouts, no early failures). This is not necessarily accurate for all components, and certainly not for most mechanical components. In addition, certain aspects of reliability analysis, such as preventive maintenance analysis and burn-in analysis, cannot be performed on components/subsystems that follow the exponential distribution.

So, the basic negative aspect of this approach of reliability prediction is that it is not reflective of the product's actual performance. Therefore, reliability prediction results may be very different from field results, and, as a final result, the reliability prediction will be unsuccessful.

1.1.5 Reliability Prediction for Mean Time Between Failures

Reliability prediction tools such as the ITEM ToolKit are essential when the reliability of electronic and mechanical components, systems, and projects is critical for life safety. Certain products and systems developed for commercial, military, or other applications often need absolute ensured reliability and consistent performance. However, electronics and mechanical products, systems, and components are naturally prone to eventual breakdown owing to any number of environmental variables, such as heat, stress, moisture, and moving parts. The main question is not if there will be failures, but "When?"

Reliability is a measure of the frequency of failures over time [7].

1.1.6 About Reliability Software

The reliability software modules of the ITEM ToolKit [7] provide a user-friendly interface that allows one to construct, analyze, and display system models using the module's interactive facilities. Building hierarchies and adding new components could not be easier. ToolKit calculates the failure rates, including mean time between failures (MTBFs), associated with new components as they are added to the system, along with the overall system failure rate. Project data may be viewed via grid view and dialog view simultaneously, allowing predictions to be performed with a minimum of effort.

Each reliability prediction module is designed to analyze and calculate component, subsystem, and system failure rates in accordance with the appropriate standard. After the analysis is complete, ITEM ToolKit's integrated environment comes into its own with powerful conversion facilities to transfer data to other reliability software modules. For example, you can transfer your MIL-217 project data to FMECA or your Bellcore project to RBD. These powerful features transfer as much of the available information as possible, saving valuable time and effort [7].

The following is an interesting statement from ReliaSoft's analysis of the current situation in reliability prediction [8]:

To obtain high product reliability, consideration of reliability issues should be integrated from the very beginning of the design phase. This leads to the concept of reliability prediction. The objective of reliability prediction is not limited to predicting whether reliability goals, such as MTBF, can be reached. It can also be used for:

- Identifying potential design weaknesses
- Evaluating the feasibility of a design
- Comparing different designs and life-cycle costs
- Providing models statement from system reliability/availability analysis
- Aiding in business decisions such as budget allocation and scheduling

Once the product's prototype is available, lab tests can be utilized to obtain reliability predictions. Accurate prediction of the reliability of electronic products requires knowledge of the components, the design, the manufacturing process and the expected operating conditions. Several different approaches have been developed to achieve reliability prediction of electronic systems and components. Each approach has its advantages and disadvantages. Among these approaches, three main categories are often used within government and industry:

- empirical (standard bases);
- physics of failure, and;
- life testing.

The following provides an overview of all three approaches [8].

1.1.6.1 MIL-HDBK-217 Predictive Method

MIL-HDBK-217 is very well known in military and commercial industries. Version MIL-HDBK-217F was released in 1991 and had two revisions.

The MIL-HDBK-217 predictive method consists of two parts: one is known as the *parts count* method and the other is called the *part stress* method [8]. The parts count method assumes typical operating conditions of part complexity, ambient temperature, various electrical stresses, operation mode, and environment (called *reference conditions*). The failure rate for a part under the reference conditions is calculated as

$$\lambda_{b,i} = \sum_{i=1}^{n} (\lambda_{\text{ref}})_i$$

where λ_{ref} is the failure rate under the reference conditions and i is the number of parts.

Since the parts may not operate under the reference conditions, the real operating conditions may result in failure rates different than those given by the "parts count" method. Therefore, the part stress method requires the specific part's complexity, application stresses, environmental factors, and so on. These adjustments are called *Pi factors*. For example, MIL-HDBK-217 provides many environmental conditions, expressed as π_E, ranging from "ground benign" to "cannon launch." The standard also provides multilevel quality specifications that are expressed as π_Q. The failure rate for parts under specific operating conditions can be calculated as

$$\lambda = \sum_{i=1}^{n}(\lambda_{\text{ref},i} \times \pi_S \times \pi_T \times \pi_E \times \pi_Q \times \pi_A)$$

where π_S is the stress factor, π_T is the temperature factor, π_E is the environment factor, π_Q is the quality factor, and π_A is the adjustment factor.

1.1.6.2 Bellcore/Telcordia Predictive Method

Bellcore was a telecommunication research and development company that provided joint R&D and standards setting for AT&T and its co-owners. Bellcore was not satisfied with the military handbook methods for application with their commercial products, so Bellcore designed its own reliability prediction standard for commercial telecommunication products. Later, the company was acquired by Science Applications International Corporation (SAIC) and the company's name was changed to Telcordia. Telcordia continues to revise and update the Bellcore standard. Presently, there are two updates: SR-332 Issue 2 (September 2006) and SR-332 Issue 3 (January 2011), both titled "Reliability prediction procedure for electronic equipment."

The Bellcore/Telcordia standard assumes a serial model for electronic parts and it addresses failure rates at both the infant mortality stage and at the steady-state stages utilizing Methods I, II, and III. Method I is similar to the MIL-HDBK-217F parts count and part stress methods, providing the generic failure rates and three part stress factors: device quality factor π_Q, electrical stress factor π_S, and temperature stress factor π_T. Method II is based on combining Method I predictions with data from laboratory tests performed in accordance with specific SR-332 criteria. Method III is a statistical prediction of failure rate based on field tracking data collected in accordance with specific SR-332 criteria. In Method III, the predicted failure rate is a weighted average of the generic steady-state failure rate and the field failure rate.

1.1.6.3 Discussion of Empirical Methods

Although empirical prediction standards have been used for many years, it is wise to use them with caution. The advantages and disadvantages of empirical methods have been discussed, and a brief summary from publications in industry, military, and academia is presented next [8].

Advantages of empirical methods:

1. Easy to use, and a lot of component models exist.
2. Relatively good performance as indicators of inherent reliability.

Disadvantages of empirical methods:

1. Much of the data used by the traditional models is out of date.
2. Failure of the components is not always due to component-intrinsic mechanisms, but can be caused by the system design.
3. The reliability prediction models are based on industry-average values of failure rate, which are neither vendor specific nor device specific.
4. It is hard to collect good-quality field and manufacturing data, which are needed to define the adjustment factors, such as the Pi factors in MIL-HDBK-217.

1.1.7 Physics of Failure Methods

In contrast to empirical reliability prediction methods that are based on the statistical analysis of historical failure data, a physics of failure approach is based on the understanding of the failure mechanism and applying the physics of failure model to the data. Several popularly used models are discussed next.

1.1.7.1 Arrhenius's Law
One of the earliest acceleration models predicts how the time to failure of a system varies with temperature. This empirically based model is known as the *Arrhenius equation.* Generally, chemical reactions can be accelerated by increasing the system temperature. Since it is a chemical process, the aging of a capacitor (such as an electrolytic capacitor) is accelerated by increasing the operating temperature. The model takes the following form:

$$L(T) = A \exp\left(\frac{E_a}{kT}\right)$$

where $L(T)$ is the life characteristic related to temperature, A is a scaling factor, E_a is the activation energy, k is the Boltzmann constant, and T is the temperature.

1.1.7.2 Eyring and Other Models
While the Arrhenius model emphasizes the dependency of reactions on temperature, the Eyring model is commonly used for demonstrating the dependency of reactions on stress factors other than temperature, such as mechanical stress, humidity, or voltage. The standard equation for the Eyring model [8] is as follows:

$$L(T, S) = A T^\alpha \exp\left[\frac{E_a}{kT} + \left(B + \frac{C}{T}\right) S\right]$$

where $L(T,S)$ is the life characteristic related to temperature and another stress, A, α, B, and C are constants, S is a stress factor other than temperature, and T is absolute temperature.

According to different physics of failure mechanisms, one more factor (i.e., stress) can be either removed or added to the standard Eyring model. Several models are similar to the standard Eyring model.

Electronic devices with aluminum or aluminum alloy with small percentages of copper and silicon metallization are subject to corrosion failures and therefore can be described with the following model [8]:

$$L(\text{RH}, V, T) = B_0 \exp[(-a)\text{RH}] f(V) \exp\left(\frac{E_a}{kT}\right)$$

where B_0 is an arbitrary scale factor, α is equal to 0.1 to 0.15 per %RH, and $f(V)$ is an unknown function of applied voltage, with an empirical value of 0.12–0.15.

1.1.7.3 Hot Carrier Injection Model

Hot carrier injection describes the phenomena observed in metal–oxide–semiconductor field-effect transistors (MOSFETs) by which the carrier gains sufficient energy to be injected into the gate oxide, generate interface or bulk oxide defects, and degrade MOSFET characteristics such as threshold voltage, transconductance, and so on [8].

For n-channel devices, the model is given by

$$L(I, T) = B(I_{\text{sub}})^{-N} \exp\left(\frac{E_a}{kT}\right)$$

where B is an arbitrary scale factor, I_{sub} is the peak substrate current during stressing, N is equal to a value from 2 to 4 (typically 3), and E_a is equal to -0.1 to -0.2 eV.

For p-channel devices, the model is given by:

$$L(I, T) = B(I_{\text{gate}})^{-M} \exp\left(\frac{E_a}{kT}\right)$$

where B is an arbitrary scale factor, I_{gate} is the peak gate current during stressing, M is equal to a value from 2 to 4, and E_a is equal to -0.1 to -0.2 eV.

ReliaSoft's "Reliability prediction methods for electronic products" [8] states:

> Since electronic products usually have a long time period of useful life (i.e., the constant line of the bathtub curve) and can often be modeled using an exponential distribution, the life characteristics in the above physics of failure models can be replaced by MTBF (i.e., the life characteristic in the exponential distribution). However, if you think your products do not exhibit a constant failure rate and therefore cannot be described by an exponential distribution, the life characteristic usually will not be the MTBF. For example, for the Weibull distribution, the life characteristic is the scale parameter *eta* and for the lognormal distribution, it is the *log mean*.

1.1.7.4 Black Model for Electromigration

Electromigration is a failure mechanism that results from the transfer of momentum from the electrons, which move in the applied electric field, to the ions, which make up the lattice of the interconnect material. The most common failure mode is "conductor open." With the decreased structure of integrated circuits (ICs), the increased current density makes this failure mechanism very important in IC reliability.

At the end of the 1960s, J. R. Black developed an empirical model to estimate the mean time to failure (MTTF) of a wire, taking electromigration into consideration, which is now generally known as the *Black model*. The Black model employs external heating and increased current density and is given by

$$\text{MTTF} = A_0 (J - J_{\text{threshold}})^{-N} \exp\left(\frac{E_a}{kT}\right)$$

where A_0 is a constant based on the cross-sectional area of the interconnect, J is the current density, $J_{\text{threshold}}$ is the threshold current density, E_a is the activation energy, k is the Boltzmann constant, T is the temperature, and N is a scaling factor.

The current density J and temperature T are factors in the design process that affect electromigration. Numerous experiments with different stress conditions have been reported in the literature, where the values have been reported in the range between 2 and 3.3 for N, and 0.5 to 1.1 eV for E_a. Usually, the lower the values, the more conservative the estimation.

1.1.7.5 Discussion of Physics of Failure Methods

A given electronic component will have multiple failure modes, and the component's failure rate is equal to the sum of the failure rates of all modes (i.e., humidity, voltage, temperature, thermal cycling, and so on). The authors of this method propose that the system's failure rate is equal to the sum of the failure rates of the components involved. In using the aforementioned models, the model parameters can be determined from the design specifications or operating conditions. If the parameters cannot be determined without conducting a test, the failure data obtained from the test can be used to get the model parameters. Software products such as ReliaSoft's ALTA can help analyze the failure data; for example, to analyze the Arrhenius model. For this example, the life of an electronic component is considered to be affected by temperature. The component is tested under temperatures of 406, 416, and 426 K. The usage temperature level is 400 K. The Arrhenius model and the Weibull distribution are used to analyze the failure data in ALTA.

Advantages of physics of failure methods:

1. Modeling of potential failure mechanisms based on the physics of failure.
2. During the design process, the variability of each design parameter can be determined.

Disadvantages of physics of failure methods:

1. The testing conditions do not accurately simulate the field conditions.
2. There is a need for detailed component manufacturing information, such as material, process, and design data.
3. Analysis is complex and could be costly to apply.
4. It is difficult (almost impossible) to assess the entire system.

Owing to these limitations, this is not generally a practical methodology.

1.1.8 Life Testing Method

As mentioned earlier, time-to-failure data from life testing may be incorporated into some of the empirical prediction standards (i.e., Bellcore/Telcordia Method II) and may also be necessary to estimate the parameters for some of the physics of failure models. But the term *life testing method* should refer specifically to a third type of approach for predicting the reliability of electronic products. With this method, a test is conducted on a sufficiently large sample of units operating under normal usage conditions. Times to failure are recorded and then analyzed with an appropriate statistical distribution in order to estimate reliability metrics such as the B10 life. This type of analysis is often referred to as *life data analysis* or *Weibull analysis.*

ReliaSoft's Weibull++ software is a tool for conducting life data analysis. As an example, suppose that an IC board is tested in the lab and the failure data are recorded. But failure data during long period of use cannot obtained, because accelerated life testing (ALT) methods are not based on accurate simulation of the field conditions.

1.1.8.1 Conclusions

In the ReliaSoft article [8], three approaches for electronic reliability prediction were discussed. The empirical (or standards based) method, which is close to the theoretical approach in practical usage, can be used in the pre-design stage to quickly obtain a rough estimation of product reliability. The physics of failure and life testing methods can be used in both design and production stages. When using the physics of failure approach, the model parameters can be determined from design specifications or from test data. But when employing the life testing method, since the failure data, the *prediction results usually are not more accurate than those from a general standard model.*

For these reasons, the traditional approaches to reliability prediction are often unsuccessful when used in industrial applications.

And one more important reason is these approaches are not closely connected with the system of obtaining accurate initial information for calculating reliability prediction during any period of use.

Some of the topics covered in the ANSI/VITA 51.2 standard [9] include reliability mathematics, organization and analysis of data, reliability modeling, and system reliability evaluation techniques. Environmental factors and stresses are taken into account in computing the reliability of the components involved. The limitations of models, methods, procedures, algorithms, and programs are outlined. The treatment of maintained systems is designed to aid workers in analyzing systems with more realistic assumptions. FTA, including the most recent developments, is also extensively discussed. Examples and illustrations are included to assist the reader in solving problems in their own area of practice. These chapters provide a guided presentation of the subject matter, addressing both the difficulties expected for the beginner, while addressing the needs of the more experienced reader.

Failures have been a problem since the very first computer. Components burned out, circuits shorted or opened, solder joints failed, pins were bent, and metals reacted unfavorably when joined. These and countless other failure mechanisms have plagued the computer industry from the very first circuits to today.

As a result, computer manufacturers realize that reliability predictions are very important to the management of their product's profitability and life cycle. They employ these predictions for a variety of reasons, including those detailed in ANSI/VITA 51.2 [9]:

- Helping to assess the effect of product reliability on the maintenance activity and quantity of spare units required for acceptable field performance of any particular system. Reliability prediction can be used to establish the number of spares needed and predict the frequency of expected unit level maintenance.
- Reliability prediction reasons.
  - Prediction of the reliability of electronic products requires knowledge of the components, the design, the manufacturing process, and the expected operating conditions. Once the prototype of a product is available, tests can be utilized to obtain reliability predictions. Several different approaches have been developed to predict the reliability of electronic systems and components. Among these approaches, three main categories are often used within government and industry: empirical (standards based), physics of failure, and life testing.
  - Empirical prediction methods are based on models developed from statistical curve fitting of historical failure data, which may have been collected in the field, in-house, or from manufacturers. These methods tend to present reasonable estimates of reliability for similar or slightly modified parts. Some parameters in the curve function can be modified by integrating existing engineering knowledge. The assumption is made that system or equipment failure causes are inherently linked to components

whose failures are essentially independent of each other. There are many different empirical methods that have been created for specific applications.

– The physics of failure approach is based on the understanding of the failure mechanism and applying the physics of failure model to the data. Physics of failure analysis is a methodology of identifying and characterizing the physical processes and mechanisms that cause failures in electronic components. Computer models integrating deterministic formulas from physics and chemistry are the foundation of physics of failure.

While these traditional approaches provide good theoretical approaches, they are unable to reflect or account for actual reliability changes that occur during service life when usage interaction, and the effects of real-world input, influences the product's reliability. For these reasons, it also is often not successful.

1.1.8.2 Failure of the Old Methods

Today, we find that the old methods of predicting reliability in electronics have begun to fail us. MIL-HDBK-217 has been the cornerstone of reliability prediction for decades. But MIL-HDBK-217 is rapidly becoming irrelevant and unreliable as we venture into the realm of nanometer geometry semiconductors and their failure modes. The uncertainty of the future of long established methods has many in the industry seeking alternative methods.

At the same time, on the component supplier side, semiconductor suppliers have been able to provide such substantial increases in component reliability and operational lifetimes that they are slowly beginning to drop MIL-STD-883B testing, and many have dropped their lines of mil-spec parts. A major reason contributing to this is that instead of focusing on mil-spec parts they have moved their focus to commercial-grade parts, where the unit volumes are much higher. In recent times the purchasing power of military markets has diminished to the point where they no longer have the dominant presence and leverage. Instead, system builders took their commercial-grade devices, sent them out to testing labs, and found that the majority of them would, in fact, operate reliably at the extended temperature ranges and environmental conditions required by the mil-spec. In addition, field data gathered over the years has improved much of the empirical data necessary for complex algorithms for reliability prediction [10–14].

The European Power Supply Manufacturers Association [15] provides engineers, operations managers, and applied statisticians with both qualitative and quantitative tools for solving a variety of complex, real-world reliability problems. There is wealth of accompanying examples and case studies [15]:

• Comprehensive coverage of assessment, prediction, and improvement at each stage of a product's life cycle.

- Clear explanations of modeling and analysis for hardware ranging from a single part to whole systems.
- Thorough coverage of test design and statistical analysis of reliability data.
- A special chapter on software reliability.
- Coverage of effective management of reliability, product support, testing, pricing, and related topics.
- Lists of sources for technical information, data, and computer programs.
- Hundreds of graphs, charts, and tables, as well as over 500 references.
- PowerPoint slides which are available from the Wiley editorial department.

Gipper [16] provides a comprehensive overview of both qualitative and quantitative aspects of reliability. Mathematical and statistical concepts related to reliability modeling and analysis are presented along with an important bibliography and a listing of resources, including journals, reliability standards, other publications, and databases. The coverage of individual topics is not always deep, but should provide a valuable reference for engineers or statistical professionals working in reliability.

There are many other publications (mostly articles and papers) that relate to the current situation in the methodological aspects of reliability prediction. In the *Reliability and Maintainability Symposiums Proceedings* (RAMS) alone there have been more than 100 papers published in this area. For example, RAMS 2012 published six papers. Most of them related to reliability prediction methods in software design and development.

Both physics-based modeling and simulation and empirical reliability have been subjects of much interest in computer graphics.

The following provides the basic content of the abstracts of some of the articles in reliability prediction from the RAMS.

1. Cai *et al.* [17] present a novel method of field reliability prediction considering environment variation and product individual dispersion. Wiener diffusion process with drift was used for degradation modeling, and a link function which presents degradation rate is introduced to model the impact of varied environments and individual dispersion. Gamma, transformed-Gamma (T-Gamma), and the normal distribution with different parameters are employed to model right-skewed, left-skewed, and symmetric stress distribution in the study case. Results show obvious differences in reliability, failure intensity, and failure rate compared with a constant stress situation and with each other. The authors indicate that properly modeled (proper distribution type and parameters) environmental stress is useful for varied environment oriented reliability prediction.

2. Chigurupati *et al.* [18] explore the predictive abilities of a machine learning technique to improve upon the ability to predict individual component times until failure in advance of actual failure. Once failure is predicted, an impending problem can be fixed before it actually occurs. The developed

algorithm was able to monitor the health of 14 hardware samples and notify us of an impending failure providing time to fix the problem before actual failure occurred.

3. Wang *et al.* [19] deals with the concept of space radiation environment reliability for satellites and establishes a model of space radiation environment reliability prediction, which establishes the relationship among system failure rate and space radiation environment failure rate and nonspace radiation environment failure rate. It provides a method of space radiation environment reliability prediction from three aspects:

(1) Incorporating the space radiation environment reliability into traditional reliability prediction methods, such as FIDES and MIL-HDBK-217.

(2) Summing up the total space radiation environment reliability failure rate by summing the total hard failure rate and soft failure rate of the independent failure rates of SEE, total ionizing dose (TID), and displacement damage (DD).

(3) Transferring TID/DD effects into equivalent failure rate and considering single event effects by failure mechanism in the operational conditions of duty hours within calendar year. A prediction application case study has been illustrated for a small payload. The models and methods of space radiation environment reliability prediction are used with ground test data of TID and single event effects for field programmable gate arrays.

4. In order to utilize the degradation data from hierarchical structure appropriately, Wang *et al.* first collected and classified feasible degradation data from a system and subsystems [20]. Second, they introduced the support vector machine method to model the relationship among hierarchical degradation data, and then all degradation data from subsystems are integrated and transformed to the system degradation data. Third, with this processed information, a prediction method based on Bayesian theory was proposed, and the hierarchical product's lifetime was obtained. Finally, an energy system was taken as an example to explain and verify the method in this paper; the method is also suitable for other products.

5. Jakob *et al.* used knowledge about the occurrence of failures and knowledge of the reliability in different design stages [21]. To show the potential of this approach, the paper presents an application for an electronic braking system. In a first step, the approach presented shows investigations of the physics of failure based on the corresponding acceleration model. With knowledge of the acceleration factors, the reliability can be determined for each component of the system for each design stage. If the failure mechanisms occurring are the same for each design stage, the determined reliability of earlier design stages could be used as pre-knowledge as well. For the calculation of the system reliability, the reliability values of all

system components are brought together. In this paper, the influence of sample size, stress levels of testing, and test duration on reliability characteristics are investigated. As already noted, an electronic braking system served here as the system to be investigated. Accelerated testing was applied (i.e., testing at elevated stress levels). If pre-knowledge is applied and the same test duration is observed, this allows for the conclusion of higher reliability levels. Alternatively, the sample size can be reduced compared with the determination of the reliability without the usage of pre-knowledge. It is shown that the approach presented is suitable to determine the reliability of a complex system such as an electronic braking system. An important part for the determination of the acceleration factor (i.e., the ratio of the lifetime under use conditions to the lifetime under test conditions) is the knowledge of the exact field conditions. Because that is difficult in many cases, further investigations are deemed necessary.

6. Today's complex designs, which have intricate interfaces and boundaries, cannot rely on the MIL-HDBK-217 methods to predict reliability. Kanapady and Adib [22] present a superior reliability prediction approach for design and development of projects that demand high reliability where the traditional prediction approach has failed to do the job. The reliability of a solder ball was predicted. Sensitivity analysis, which determines factors that can mitigate or eliminate the failure mode(s), was performed. Probabilistic analysis, such as the burden capability method, was employed to assess the probability of failure mode occurrences, which provides a structured approach to ranking of the failure modes, based on a combination of their probability of occurrence and severity of their effects.

7. Microelectronics device reliability has been improving with every generation of technology, whereas the density of the circuits continues to double approximately every 18 months. Hava *et al.* [23] studied field data gathered from a large fleet of mobile communications products that were deployed over a period of 8 years in order to examine the reliability trend in the field. They extrapolated the expected failure rate for a series of microprocessors and found a significant trend whereby the circuit failure rate increases approximately half the rate of the technology, going up by approximately $\sqrt{2}$ in that same 18-month period.

8. Thaduri *et al.* [24] studied the introduction, functioning, and importance of a constant fraction discriminator in the nuclear field. Furthermore, the reliability and degradation mechanisms that affect the performance of the output pulse with temperature and dose rates act as input characteristics was properly explained. Accelerated testing was carried out to define the life testing of the component with respect to degradation in output transistor–transistor logic pulse amplitude. Time to failure was to be properly quantified and modeled accordingly.

9. Thaduri *et al.* [25] also discussed several reliability prediction models for electronic components, and comparison of these methods was also illustrated. A combined methodology for comparing the cost incurred for prediction was designed and implemented with an instrumentation amplifier and a bipolar junction transistor (BJT). By using the physics of failure approach, the dominant stress parameters were selected on the basis of a research study and were subjected to both an instrumentation amplifier and a BJT. The procedure was implemented using the methodology specified in this paper and modeled the performance parameters accordingly. From the prescribed failure criteria, the MTTF was calculated for both the components. Similarly, using the 217Plus reliability prediction book, the MTTF was also calculated and compared with the prediction using physics of failure. Then, the costing implications of both the components were discussed and compared. For critical components like an instrumentation amplifier, it was concluded that though the initial cost of physics of failure prediction is too high, the total cost incurred, including the penalty costs, is lower than that of a traditional reliability prediction method. But for noncritical components like a BJT, the total cost of physics of failure approach was too high compared with a traditional approach, and hence a traditional approach was more efficient. Several other factors were also compared for both reliability prediction methods.

Much more literature on methodological approaches to reliability prediction are available.

The purpose of the MIL-HDBK-217F handbook [26] is to establish and maintain consistent and uniform methods for estimating the inherent reliability (i.e., the reliability of a mature design) of military electronic equipment and systems. It provides a common basis for reliability predictions during acquisition programs for military electronic systems and equipment. It also establishes a common basis for comparing and evaluating reliability predictions of related or competitive designs.

Another document worthy for discussion is the Telecordia document, Issue 4 of SR-332 [27]. This provides all the tools needed for predicting device and unit hardware reliability, and contains important revisions to the document. The *Telcordia Reliability Prediction Procedure* has a long and distinguished history of use both within and outside the telecommunications industry. Issue 4 of SR-332 provides the only hardware reliability prediction procedure developed from the input and participation of a cross-section of major industrial companies. This lends the procedure and the predictions derived from it a high level of credibility free from the bias of any individual supplier or service provider.

Issue 4 of SR-332 contains the following:

- Recommended methods for prediction of device and unit hardware reliability. These techniques estimate the mean failure rate in FITs for electronic equipment. This procedure also documents a recommended method for predicting serial system hardware reliability.

- Tables needed to facilitate the calculation of reliability predictions.
- Revised generic device failure rates, based mainly on new data for many components.
- An extended range of complexity for devices, and the addition of new devices.
- Revised environmental factors based on field data and experience.
- Clarification and guidance on items raised by forum participants and by frequently asked questions from users.

Lu *et al.* [28] describe an approach to real-time reliability prediction, applicable to an individual product unit operating under dynamic conditions. The concept of conditional reliability estimation is extended to real-time applications using time-series analysis techniques to bridge the gap between physical measurement and reliability prediction. The model is based on empirical measurements is, self-generating, and applicable to online applications. This approach has been demonstrated at the prototype level. Physical performance is measured and forecast across time to estimate reliability. Time-series analysis is adapted to forecast performance. Exponential smoothing with a linear level and trend adaptation is applied. This procedure is computationally recursive and provides short-term, real-time performance forecasts that are linked directly to conditional reliability estimates. Failure clues must be present in the physical signals, and failure must be defined in terms of physical measures to accomplish this linkage. On-line, real-time applications of performance reliability prediction could be useful in operational control as well as predictive maintenance.

Section 8.1 in Chapter 8 of the *Engineering Statistics e-Handbook* [29] considers lifetime or repair models. As seen earlier, repairable and nonrepairable reliability population models have one or more unknown parameters. The classical statistical approach considers these parameters as fixed but unknown constants to be estimated (i.e., "guessed at") using sample data taken randomly from the population of interest. A confidence interval for an unknown parameter is really a frequency statement about the likelihood that the numbers calculated from a sample capture the true parameter. Strictly speaking, one cannot make probability statements about the true parameter since it is fixed, not random. The Bayesian approach, on the other hand, treats these population model parameters as random, not fixed, quantities. Before looking at the current data, it uses old information, or even subjective judgments, to construct a prior distribution model for these parameters. This model expresses our starting assessment about how likely various values of the unknown parameters are. We then make use of the current data (via *Bayes' formula*) to revise this starting assessment, deriving what is called the posterior distribution model for the population model parameters. Parameter estimates, along with confidence intervals (known as credibility intervals), are calculated directly from the posterior distribution. Credibility intervals are legitimate probability statements about the unknown parameters, since these parameters now are considered random, not fixed.

Unfortunately, it is unlikely in most applications that data will ever exist to validate a chosen prior distribution model. But parametric Bayesian prior models are often chosen because of their flexibility and mathematical convenience. In particular, conjugate priors are a natural and popular choice of Bayesian prior distribution models.

1.1.9   Section Summary

1. Section 1.1 concludes that most of the approaches considered are difficult to use successfully in practice.
2. The basic cause for this is a lack of close connection to the source, which is necessary for obtaining accurate initial information that is needed for calculating changing reliability parameters during the product's life cycle.
3. Three basic methods of reliability prediction were discussed:
   - Empirical reliability prediction methods that are based on the statistical analysis of historical failure data models, developed from statistical curves. These methods are not considered accurate simulations of field situations and are an obstacle to obtaining accurate initial information for calculating the dynamics of changing failure (degradation) parameters during a product or technology service life.
   - Physics of failure approach, which is based on the understanding of the failure mechanism and applying the physics of failure model to the data. However, one cannot obtain data for service life during the design and manufacturing stages of a new model of product/technology. Accurate initial information from the field during service life is not available during these stages of development.
   - Laboratory-based or proving ground based life testing reliability prediction, which uses ALT in the laboratory but does not accurately simulate changing parameters encountered in the field during service life of the product/technology.
4. Recalls, complaints, injuries and deaths, and significant costs are direct results of these prediction failures.
5. Real products rarely exhibit a constant failure rate and, therefore, cannot be accurately described by exponential, lognormal, or other theoretical distributions. Real-life failure rates are mostly random.
6. Reliability prediction is often considered as a separate issue, but, in real life, reliability prediction is an essential interacted element of a product/technology performance prediction [30].

Analysis demonstrates that the current status of product/technology methodological aspects of reliability prediction for industries such as electronic, automotive, aircraft, aerospace, off-highway, farm machinery, and others is not very successful. The basic cause is the difficulty of obtaining accurate initial information for specific product prediction calculations during the real-world use of the product.

Accurate prediction requires information similar to that experienced in the real world.

There are many publications on the methodological aspects of reliability prediction that commonly have similar problems; for example, see Refs [31–40].

1.2 Current Situation in Practical Reliability Prediction

There are far fewer publications that relate to the practical aspects of reliability prediction. Sometimes authors call their publications by some name indicating practical reliability prediction, but the approaches they present are not generally useful in a practical way.

The electronics community is comprised of representatives from electronics suppliers, system integrators, and the Department of Defense (DoD). The majority of the reliability work is driven by the user community that depends heavily on solid reliability data. BAE Systems, Bechtel, Boeing, General Dynamics, Harris, Lockheed Martin, Honeywell, Northrop Grumman, and Raytheon are some of the better known demand-side contributors to the work done by the reliability community. These members have developed consensus documents to define electronics failure rate prediction methodologies and standards. Their efforts have produced a series of documents that have been ANSI and VITA ratified. In some cases, these standards provide adjustment factors to existing standards.

The reliability community addresses some of the limitations of traditional prediction practices with a series of subsidiary specifications that contain "best practices" within the industry for performing electronics failure rate predictions. The members recognize that there are many industry reliability methods, each with a custodian and a methodology of acceptable practices, to calculate electronics failure rate predictions. If additional standards are required, for use by electronics module suppliers, a new subsidiary specification will be considered by the working group.

ANSI/VITA 51.3 *Qualification and Environmental Stress Screening in Support of Reliability Predictions* provides information on how qualification levels and environmental stress screening may be used to influence reliability.

Although they sometimes call what are essentially empirical prediction methods "practical methods", these approaches do not provide accurate and successful prediction during the product's service life. Let us briefly review these and some other relevant publications.

David Nicholls provides an overview in this area [41]:

- For more than 25 years, there has been a passionate dialog throughout the reliability engineering community as to the appropriate use of empirical and physics-based reliability models and their associated benefits, limitations, and risks.
- Over its history, the Reliability Information Analysis Center has been intimately involved in this debate. It has developed models for MIL-HDBK-217, as well as alternative empirically based methodologies such as 217Plus. It has published books related to physics-of-failure modeling approaches and developed the Web-Accessible Repository for Physics-based Models to support the use of physics-of-failure approaches. In DoD-sponsored documents it also had published ideal attributes to identify future reliability predicting methods.
- Empirical predicting methods, particularly those based on field data, provide average failure rates based on the complexities of actual environmental and operational stresses. Generally, they should be used only in the absence of actual relevant comparison data. They are frequently criticized, however, for being grossly inaccurate in prediction of actual field failure rates, and for being used and accepted primarily for verification of contractual reliability requirements.
- What should your response be (if protocol allows) if you are prohibited from using one prediction approach over another because it is claimed that, historically, those prediction methods are "inaccurate," or they are deemed too labor intensive/cost ineffective to perform based on past experience?
- Note that the terms "predicting," "assessment," and "estimation" are not always distinguishable in the literature.
- A significant mistake typically made in comparing actual field reliability data with the original prediction is that the connection is lost between the root failure causes experienced in the field and the intended purpose/coverage of the reliability prediction approach. For example, MIL-HDBK-217 addresses only electronic and electromechanical components. Field failures whose root failure cause is traced to mechanical components, software, manufacturing deficiencies, and so on, should not be scored against a MIL-HDBK-217 prediction.
- Data that form the basis for either an empirical or physics-based prediction has various factors that will influence any uncertainty in the assessments made using the data. One of the most important factors is relevancy. Relevancy is defined as the similarity between the predicted and fielded product/system architectures, complexities, technologies, and environmental/operational stresses.
- Empirical reliability prediction models based on field data inherently address all of the failure mechanisms associated with reliability in the field, either explicitly (e.g., factors based on temperature) or implicitly

(generic environmental factors that "cover" all of the failure mechanisms not addressed explicitly). They do not, however, assess or consider the impact of these various mechanisms on specific failure modes, nor do they (generally) consider "end-of-life" issues.

- The additional insight required to ensure an objective interpretation of results for physics-based prediction includes:
  - What failure mechanisms/modes of interest are addressed by the predicting?
  - What are the reliability or safety risks associated with the failure mechanisms/modes not addressed by the prediction?
  - Does the prediction consider interactions between failure mechanisms (e.g., combined maximum vibration level and minimum temperature level).

As a result of the overview, Nicholls concluded:

1. As the reader may have noticed, there was one question brought up in the paper that was never addressed: "How can one ensure that prediction results will not be misinterpreted or misapplied, even though all assumptions and rationale have been meticulously documented and clearly stated?"
2. Unfortunately, the answer is: "You can't." Regardless of the care taken to ensure a technically valid and supportable analysis, empirical and physics-based predicting will always need to be justified as to why the predicted reliability does not reflect the measured reliability in the field.

From the aforementioned information we can conclude that current prediction methods in engineering are primarily related to computer science, and even there they are not entirely successful. Given this fact, how much less is known about prediction, especially accurate prediction, that relates to automotive, aerospace, aircraft, off-highway, farm machinery, and others? We need to conclude that it is even less developed in these areas.

A fatigue life prediction method for practical engineering use was proposed by Theil [42]:

- According to this method, the influence of overload events can be taken in account.
- The validation was done using uniaxial tests carried out on metallic specimens.

It is generally accepted that during a product's service life high load cycles will occur in addition to the normally encountered operational loads. Therefore, the development of an accurate fatigue life prediction rule which takes into account overloads in the vicinity of and slightly above yield strength, with a minimal level of effort for use in practical engineering at the design stress level, would still be highly significant.

Theil's paper [42] presents a fatigue life prediction method based on an S/N curve for constant-amplitude loading. Similarities and differences between the proposed method and the linear cumulative damage rule of Pålmgren–Miner are briefly discussed. Using the method presented, an interpretation of the Pålmgren–Miner rule from the physical point of view is given and clarified with the aid of a practical two-block loading example problem.

Unfortunately however, statistical reliability predictions rarely correlate with field performance. The basic cause is that they are based on testing methods that incorrectly simulate the real-world interacted conditions.

Therefore, similar to the contents of Section 1.1, current practical reliability prediction approaches cannot provide industry with with the necessary or appropriate tools to dramatically increase reliability, eliminate (or dramatically reduce) recalls, complaints, costs, and other technical and economic aspects of improved product performance.

1.3 From History of Reliability Prediction Development

The term "reliability prediction" has historically been used to denote the process of applying mathematical models and data for the purpose of estimating the field reliability of a system before empirical data are available for the system [43, 44].

Jones [45] considered information on the history of reliability prediction allowing his work to be placed in context with general developments in the field. This includes the development of statistical models for lifetime prediction using early life data (i.e., prognostics), the use of nonconstant failure rates for reliability prediction, the use of neural networks for reliability prediction, the use of artificial intelligence systems to support reliability engineers' decision-making, the use of a holistic approach to reliability, the use of complex discrete events simulation to model equipment availability, the demonstration of the weaknesses of classical reliability prediction, an understanding of the basic behavior of no fault founds, the development of a parametric drift model, the identification of the use of a reliability database to improve the reliability of systems, and an understanding of the issues that surround the use of new reliability metrics in the aerospace industry.

During World War II, electronic tubes were by far the most unreliable component used in electronic systems. This observation led to various studies and ad hoc groups whose purpose was to identify ways that electronic tube reliability, and the reliability of the systems in which they operated, could be improved. One group in the early 1950s concluded that:

1. There needed to be better reliability-data collected from the field.
2. Better components needed to be developed.

3. Quantitative reliability requirements needed to be established.
4. Reliability needed to be verified by test before full-scale production.

The specification requirement to have quantitative reliability requirements in turn led to the need to have a means of estimating reliability before the equipment is built and tested so that the probability of achieving its reliability goal could be estimated. This was the beginning of reliability prediction.

Then, in the 1960s, the first version of US MH-217 was published by the US Navy. This document became the standard by which reliability predictions were performed. Other sources of failure rates prediction gradually disappeared. These early sources of failure-rate predicting often included design guidance on the reliable application of electronic components.

In the early 1970s, the responsibility for preparing MH-217 was transferred to the Rome Air Development Center, who published revision B in 1974. While this MH-217 update reflected the technology at that time, there were few other efforts to change the manner in which predicting was performed. And these efforts were criticized by the user community as being too complex, too costly, and unrealistic.

While MH-217 was updated several times, in the 1980s other agencies were developing reliability predicting models unique to their industries. For example, the SAE Reliability Standard Committee developed a set of models specific to automotive electronics. The SAE committee did this because it was their belief that there were no existing prediction methodologies that applied to the specific quality levels and environments appropriate for automotive applications.

The Bellcore reliability-prediction standard is another example of a specific industry developing methodologies for their unique conditions and equipment. But regardless of the developed methodology, the conflict between the usability of a model and its accuracy has always been a difficult compromise.

In the 1990s, much of the literature on reliability prediction centered around whether the reliability discipline should focus on physics-of-failure-based or empirically based models (such as MH-217) for the qualification of reliability.

Another key development in reliability prediction related to the effects of acquisition reform, which overhauled the military standardization process. These reforms in turn led to a list of standardization documents that required priority action, because they were identified as barriers to commercial acquisition processes, as well as major cost drivers in defense acquisitions.

The premise of traditional methods, such as MH-217, is that the failure rate is primarily determined by components comprising the system. The prediction methodologies that were developed toward the end of the 1990s had the following advantages:

- they used all available information to form the best estimate of field reliability;

- they were tailorable;
- they had quantifiable statistical-confidence bounds;
- they had sensitivity to the predominant system-reliability drivers.

During this time, some reliability professionals believed that reliability modeling should focus more on physics-of-failure models to replace the traditional empirical models. Physics-of-failure models attempt to model failure mechanisms deterministically, as opposed to the traditional approach of using models based on empirical data.

Physics-of-failure techniques can be effective for estimating lifetimes due to specific failure mechanisms. These techniques are useful for ensuring that there are no quantifiable failure mechanisms that will occur in a given time period. However, many of the arguments that physics-of-failure proponents use are based on erroneous assumptions regarding empirically based reliability prediction. The fact that a failure rate can be predicted for a given part under a specific set of conditions does not imply that a failure rate is an inherent quality of a part.

Rather, the probability of failure is a complex interaction between the defect density, defect severity, and stresses incurred in operation. Failure rates predicted using empirical models are therefore typical failure rates and represent typical defect rates, design, and use conditions.

Therefore, there is a tradeoff between the model's usability and its required level of detailed data. This highlights the fact that the purpose of a reliability prediction must be clearly understood before a methodology is chosen.

From the foregoing we can conclude that current prediction methods in engineering mostly relate to computer science, and even they are *not very successful*. And as relates to prediction (especially successful prediction) of automotive, aircraft, off-highway, farm machinery, and others, we need to accept that prediction methods are even less developed.

A similar conclusion was made by Wong [46]:

> Inaccurate reliability predictions could lead to disasters such as in the case of the U.S. Space Shuttle failure. The question is: 'what is wrong with the existing reliability prediction methods?' This paper examines the methods for predicting reliability of electronics. Based on information in the literature the measured vs predicted reliability could be as far apart as five to twenty times. Reliability calculated using the five most commonly used handbooks showed that there could be a 100 times variation. The root cause for the prediction inaccuracy is that many of the first-order effect factors are not explicitly included in the prediction methods. These factors include thermal cycling, temperature change rate, mechanical shock, vibration, power on/off, supplier quality difference, reliability improvement with respect to calendar years, and

ageing. As indicated in the data provided in this paper, any one of these factors neglected could cause a variation in the predicted reliability by several times. The reliability vs ageing-hour curve showed that there was a 10 times change in reliability from 1000 ageing-hours to 10,000 ageing-hours. Therefore, in order to increase the accuracy of reliability prediction the factors must be incorporated into the prediction methods.

1.4 Why Reliability Prediction is Not Effectively Utilized in Industry

As one can see from Figure 1.1, reliability is a result of many interacting components which influence the product's performance. Therefore, one needs to understand that, if we consider reliability separately, it is different from that in the real-life situation. Reliability prediction is only one aspect that is a result of the interacted product performance in real life (i.e., only one step of product/technology performance). Klyatis [30] gives full consideration of product performance prediction.

As was previously discussed, there are many methodological approaches to different aspects of engineering prediction; these include Refs [47–59], but there are many other publications on the subject.

The problems remain:

- How can one obtain common methodological (strategic and tactical) aspects of interacted components performance for successful prediction? Reliability is only one of many interacted factors of a product's or a technology's performance (Figure 1.1).
- How does one obtain accurate initial information necessary for each particular product's successful performance prediction, including reliability, safety, durability, life cycle cost, and others?

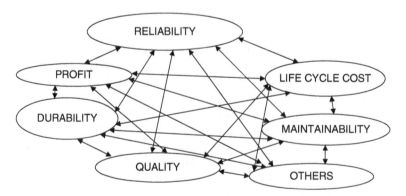

Figure 1.1 Reliability as one from interacted performance components in the real world.

In order to solve these problems, one needs to understand the basic causes of why the approaches used until now cannot solve problems of successfully predicting reliability.

One frequently encountered cause is that the management structure of many large and middle-size companies is hierarchical, with managerial responsibilities that provide no, or inadequate, interconnections or interactions with other areas of the organization, or with outside entities supplying components or services to their companies.

Too often this leads to parochial thinking and inadequate consideration of how other sections of their organization influence their product's effectiveness. Figure 1.2 demonstrates sectors of responsibility of four vice presidents (1, 2, 3, 4) in a large organization. Each vice-president's sector of responsibility consists of some subsectors, shown as subsectors 3a, 3b, and 3c, each with a corresponding responsible technical director, or some similar title (example in Figure 1.2). These subsector directors do not communicate enough with each other, because each sub-director's responsibility relates to their particular subsector only. But their work interacts and influences the final reliability of the product.

Let us consider some examples from author's practice. After reading author's publications, Mr. Takashi Shibayama, Vice President of Jatco Ltd (Japan) (design and manufacturing for automobile transmissions), indicated that he was very interested in implementing these practices at Jatco Ltd. Mr. Shibayama came to the SAE World Congress bringing his copy of author's book *Accelerated Reliability and Durability Testing Technology*, published by Wiley, which he had studied. After discussion with Dr. Klyatis, Mr. Shibayama

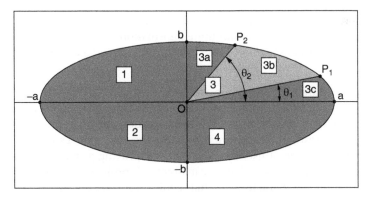

Figure 1.2 Common scheme of company's vice-president's sectors of responsibility. 1: one vice-president's area; 2: second vice-president's area; 3: third vice-president's area; 4: fourth vice-president's area; 3a: area of responsibilities of director of first department; 3b: area of responsibilities of director of second department; 3c: area of responsibilities of director of third department.

and Jatco's engineers and managers returned to Japan and asked by e-mail if I could meet with Nissan management. It was agreed that a meeting would be scheduled during the next SAE World Congress in Detroit. This meeting would be attended by leaders in the Powertrain Engineering Division at Nissan. Mr. Shibayama indicated these people would all be the expert leaders in Nissan reporting directly to the highest levels in Nissan.

One participant was from the Engine Department, and another from the Drive Train Department, one was the engineering director involved directly in human factor problems, and another the engineering director in engine problems. They first attended author's presentation at the SAE 2013 World Congress, Technical Session IDM300 Trends in Development Accelerated Reliability and Durability Testing, as well as author's "Chat with the Expert" session. The company's personnel had studied his book, *Accelerated Reliability and Durability Testing Technology*, prior to the meeting. During the meeting, they asked if he would help them to improve each of their separate areas for reliability, and he replied that he could not. The reason for this answer was that each of them was responsible for their specific area and they did not interact with the other areas of Nissan's engineering. He said that all sectors and components of the vehicle interact and are interconnected, and if we consider each component separately then you could not hope to obtain a successful improvement in prediction. He pointed out that page 65 in his book emphasized the need for solving reliability issues in a complex product, and that as long as each area of the business worked independently, overall product reliability prediction could not be improved.

From discussions with other industrial companies' managements, as well as reviewing their work, it became evident that similar situations are prevalent in other areas of industry, including, for example, the aircraft industry.

Figure 1.3 is provided as an aid in understanding why the current discussions in Sections 1.1 and 1.2 fail to provide accurate information for successful reliability prediction. A short description of the basic causes depicted in Figure 1.3 follows:

1. Reliability and durability evaluation results are usually provided directly after stress testing either in the laboratory or from proving grounds testing. This evaluation only relates to the test conditions (the laboratory or proving ground testing). But in order to know the test subject's reliability or durability in actual use, the laboratory or proving ground testing results are not adequate, as field use normally entails probabilistic, random (often nonstationary) characteristics that require more complicated methods than evaluation based on laboratory (with simple simulation of the field) or proving ground test protocols. But, this more complex methodology is seldom used. As a result, simpler evaluation methods are used, but these are not reliable predictors.

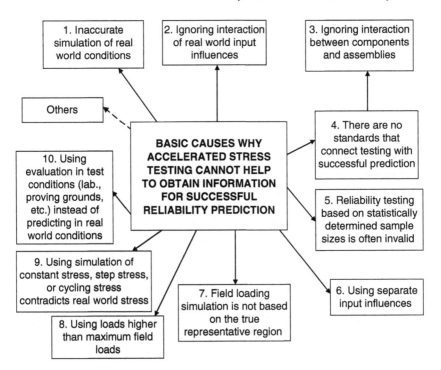

Figure 1.3 The reasons why accelerated stress testing cannot provide information for successful reliability prediction.

2. Typical testing is performed for several separate parameters, such as vibration or temperature, with allowances possibly including a few others. But in real-world operation there are many other factors influencing or affecting the test subject. Temperature, humidity, air pollution (mechanical and chemical), radiation (visible, ultraviolet, and infrared), air fluctuation, features of the road (type of road, profile, density, etc.), speed of movement, input voltage, electrostatic discharge, and other factors may be present in different combinations. If laboratory or proving grounds simulations use only several of the multi-environmental real-world input influences described here, this ignores what may well be significant interactions between different factors that will influence reliability and durability. This is one reason why current accelerated stress testing fails to accurately predict the reliability and durability of the product, whether it be components or the completed product in the field.

3. Often, companies' suppliers use only components (details) and assemblies (units) stress testing, ignoring the interconnections between these

components and assemblies. Therefore, the stress testing results are different from the test subject's results in the real world. These results cannot help to predict accurately the reliability and durability of the product in the field.

4. A basic reason for many product recalls is inaccurate simulation and mistakes in using stress testing results, which lead to inaccurate prediction of a product's reliability. The result is that an organization's profit is reduced from that which was projected or potentially could have been achieved. As Kanapady and Adib noted [22], "Money was saved years earlier by gambling with a substandard reliability program, but the short-term gain was not a good long-term investment."

5. In an attempt to compensate for unknown field factors, companies sometimes use accelerated stress testing with loads greater than maximum field loads. This alters the physics-of-degradation or the chemistry-of-degradation process obtained in actual field use. As a result, the time and character to failures, as well as the number and cost of failures, during this testing can be different from the failures in the field situation.

6. As each failure mechanism responds to stress differently, and each component of the product has several different failure mechanisms, using accelerated test data plus a single acceleration factor can result in a time between failures (MTBF) estimate that is erroneous and misleading [37].

7. There is no standard stress stimulus portfolio. Because of the vast diversity of products, each product and program will have differences. This is often forgotten.

8. Reliability assumptions based on statistically determined sample sizes are often invalid, because the samples rarely are truly random representations of production parts.

9. Often, the parameters used for field loading simulation for stress testing are not based on the truly representative region, and do not represents all areas (including climatic) of the product's use in the field.

10. Some accelerated testing uses simulation at constant stress, or step stress, or cycling stress that contradicts the real-world situation, where loading has a random stress nature.

11. As Wong noted [46], inaccurate reliability predictions could lead to disasters, such as in the case of the US Space Shuttle failure. The question is: "What is wrong with the existing reliability prediction methods?" This textbook answers this question.

These are some of the reasons, why current accelerated stress testing does not help accurately predict reliability and durability. But these limitations can be reduced or eliminated by switching to ART and ADT. Descriptions of these types of testing can be found in the literature [30, 60–66]. These publications provide guidance on how to accurately simulate the field situation for obtaining initial information for accurate reliability development and prediction.

For successful ART/ADT technology implementation, a multidisciplinary team will be needed to manage and engineer the application of this technology to a particular product. The team should include, as a minimum, the following:

- A team leader who is both a high-level manager and who also understands the strategy of this technology. The team leader must understand the principles of accurate simulation of the field situation, and what other professional disciplines need to be included in the team.
- A program manager to act as a guide throughout the process and who can remove any barriers that prevent the team from succeeding. The program manager must also be knowledgeable in the design and technology of both the product and the testing.
- Engineering resources to actually perform much of the testing. This includes selection of appropriate candidate test units, failure analysis, chemical problems solution in simulation, physical problems solution in simulation, prediction methodology, system of control development, design, diagnostic, and corrective action for mechanical, electrical, hydraulic, and so on, as well as both hardware and software development and implementation.

The team must work closely with those departments that are responsible for design, manufacturing, marketing, and selling the product.

References

1 O'Connor P, Kleyner A. (2012). *Practical Reliability Engineering*, 5th edition. John Wiley & Sons.
2 Naval Surface Warfare Center, Carderock Division. (2011). *Handbook of Reliability Prediction Procedures for Mechanical Equipment*. Logistic Technology Support, Carderockdiv, NSWC-10. West Naval Surface Warfare Center, Carderock Division, Bethesda, MD.
3 Atiyeh C, Blackwell R. (2016). Massive Takata airbag recall: everything you need to know, including full list of affected vehicles. Update: December 29. *Car and Driver*. https://blog.caranddriver.com/massive-takata-airbag-recall-everything-you-need-to-know-including-full-list-of-affected-vehicles/ (accessed January 18, 2018).
4 National Council on Measurement in Education. (2017). *Glossary of important assessment and measurement terms*. http://www.ncme.org/ncme/NCME/Resource_Center/Glossary/NCME/Resource_Center/Glossary1.aspx?hkey=4bb87415-44dc-4088-9ed9-e8515326a061#anchorR (accessed January 18, 2018).

5 Carlson NR. Miller HL Jr, Heth DS, Donahoe JW, Martin GN. (2009). *Psychology: The Science of Behaviour*, 4th Canadian edn. Pearson, Toronto.

6 Web Center for Social Research Methods. (2006). Types of reliability, in *The Research Methods Knowledge Base*. https://www.socialresearchmethods.net/kb/reltypes.php (accessed January 18, 2018).

7 Weibull.com (2005). Standards based reliability prediction: applicability and usage to augment RBDs. Part I: introduction to standards based reliability prediction and lambda predict. *Reliability HotWire* Issue 50 (April). http://www.weibull.com/hotwire/issue50/hottopics50.htm (accessed January 18, 2018).

8 ReliaSoft. Reliability prediction methods for electronic products. *Reliability EDGE* **9**(1). http://www.reliasoft.com/pubs/reliabilityedge_v9i1.pdf (accessed January 18, 2018).

9 ANSI/VITA 51.2 (2016). *Physics of Failure Reliability Predictions*.

10 Blischke WR, Prabhakar Murthy DN. (2000). *Reliability: Modeling, Prediction, and Optimization*. John Wiley & Sons.

11 Wiley Online Library. (2000). *Short Book Reviews*, **20**(3).

12 Misra KB. (1992). *Reliability Analysis and Prediction, Volume 15. A Methodology Oriented Treatment*. Elsevier Science.

13 Jones TL (2010). *Handbook of Reliability Prediction Procedures for Mechanical Equipment*. Naval Surface Warfare Center, Carderock Division.

14 An YH, Draughn RA. (1999). *Mechanical Testing of Bone–Implant Interface*. CRC Press.

15 European Power Supply Manufacturers Association. (2005). *Guidelines to understanding reliability prediction*. http://www.epsma.org/MTBF%20Report_24%20June%202005.pdf (accessed January 18, 2018).

16 Gipper J. (2014). *Choice of Reliability Prediction Methods*. http://vita.mil-embedded.com/articles/choice-reliability-prediction-methods/ (accessed February 2, 2018).

17 Cai Y-K, Wei D, Ma X-B, Zhao Y. Reliability prediction method with field environment variation. In *2015 Annual Reliability and Maintainability Symposium (RAMS)*. IEEE Press, pp. 1–7.

18 Chigurupati A, Thibaux R, Lassar N. (2016). Predicting hardware failure using machine learning. In *2016 Annual Reliability and Maintainability Symposium (RAMS)*. IEEE Press, pp. 1–6.

19 Wang Q, Chen D, Bai H. (2016). A method of space radiation environment reliability prediction. In *2016 Annual Reliability and Maintainability Symposium (RAMS)*. IEEE Press, pp. 1–6.

20 Wang L, Zhao X, Wang X, Mu M. (2016). A lifetime prediction method with hierarchical degradation data. In *2016 Annual Reliability and Maintainability Symposium (RAMS)*. IEEE Press, pp. 1–6.

21 Jakob F, Schweizer V, Bertsche B, Dobry A. (2014). Comprehensive approach for the reliability prediction of complex systems. In *2014 Reliability and Maintainability Symposium*. IEEE Press, pp. 1–6.

22 Kanapady R, Adib R. Superior reliability prediction in design and development phase. In *2013 Proceedings Annual Reliability and Maintainability Symposium (RAMS)*. IEEE Press, pp. 1–6.

23 Hava A, Qin J, Bernstein JB, Bo Y. Integrated circuit reliability prediction based on physics-of-failure models in conjunction with field study. In *2013 Proceedings Annual Reliability and Maintainability Symposium (RAMS)*. IEEE Press, pp. 1–6.

24 Thaduri A, Verma AK, Gopika V, Kumar U. Reliability prediction of constant fraction discriminator using modified PoF approach. In *2013 Proceedings Annual Reliability and Maintainability Symposium (RAMS)*. IEEE Press, pp. 1–7.

25 Thaduri A, Verma AK, Kumar U. Comparison of reliability prediction methods using life cycle cost analysis. In *2013 Proceedings Annual Reliability and Maintainability Symposium (RAMS)*. IEEE Press, pp. 1–7.

26 DAU. (1995). *MIL-HDBK-217F (Notice 2). Military Handbook: Reliability Prediction of Electronic Equipment*. Department of Defense, Washington, DC.

27 Telecordia. (2016). *SR-332, Issue 4, Reliability Prediction Procedure for Electronic Equipment*.

28 Lu H, Kolarik WJ, Lu SS. (2001). Real-time performance reliability prediction. *IEEE Transactions on Reliability* 50(4): 353–357.

29 NIST/SEMATECH. (2010). Assessing product reliability. In *Engineering Statistics e-Handbook*. US Department of Commerce, Washington, DC, chapter 8.

30 Klyatis L. (2016) *Successful Prediction of Product Performance: Quality, Reliability, Durability, Safety, Maintainability, Life-Cycle Cost, Profit, and Other Components*. SAE International, Warrendale, PA.

31 DAU. (1991). *MIL-HDBK-217F (Notice 1). Military Handbook: Reliability Prediction of Electronic Equipment*. Department of Defense, Washington, DC.

32 Telcordia. (2001). *SR-332, Issue 1, Reliability Prediction Procedure for Electronic Equipment*.

33 Telcordia. (2006). *SR-332, Issue 2, Reliability Prediction Procedure for Electronic Equipment*.

34 ITEM Software and ReliaSoft Corporation. (2015). *D490 Course Notes: Introduction to Standards Based Reliability Prediction and Lambda Predict*.

35 Foucher B, Boullie J, Meslet B, Das D. (2002). A review of reliability prediction methods for electronic devices. *Microelectronics Reliability* 42(8): 1155–1162.

36 Pecht M, Das D, Ramarkrishnan A. (2002). The IEEE standards on reliability program and reliability prediction methods for electronic equipment. *Microelectronics Reliability* **42**: 1259–1266.

37 Talmor M, Arueti S. (1997). Reliability prediction: the turnover point. In *Annual Reliability and Maintainability Symposium: 1997 Proceedings*. IEEE Press, pp. 254–262.

38 Hirschmann D, Tissen D, Schroder S, de Doncker RW. (2007). Reliability prediction for inverters in hybrid electrical vehicles. *IEEE Transactions on Power Electronics*, **22**(6): 2511–2517.

39 NIST Information Technology Library. https://www.itl.nist.gov.

40 SeMaTech International. (2000) *Semiconductor Device Reliability Failure Models*. www.sematech.org/docubase/document/3955axfr.pdf (accessed January 19, 2018).

41 Nicholls D. An objective look at predictions – ask questions, challenge answers. In *2012 Proceedings Annual Reliability and Maintainability Symposium*. IEEE Press, pp. 1–6.

42 Theil N. (2016). Fatigue life prediction method for the practical engineering use taking in account the effect of the overload blocks. *International Journal of Fatigue* **90**: 23–35.

43 Denson W. (1998). The history of reliability prediction. *IEEE Transactions on Reliability* **47**(3-SP, Part 2): SP-321–SP-328.

44 Klyatis L. The role of accurate simulation of real world conditions and ART/ADT technology for accurate efficiency predicting of the product/process. In *SAE 2014 World Congress*, paper 2014-01-0746.

45 Jones JA. (2008). Electronic reliability prediction: a study over 25 years. PhD thesis, University of Warwick.

46 Wong KL. (1990). What is wrong with the existing reliability prediction methods? *Quality and Reliability Engineering International* **6**(4): 251–257.

47 Black AI. (1989). Bellcore system hardware reliability prediction. In *Proceedings Annual Reliability and Maintainability Symposium*.

48 Bowles JB. (1992). A survey of reliability prediction procedures for microelectronic devices. *IEEE Transactions in Reliability* **41**: 2–12.

49 Chan HT, Healy JD. (1985). Bellcore reliability prediction. In *Proceedings Annual Reliability and Maintainability Symposium*.

50 Healy JD, Aridaman KJ, Bennet JM. (1999). Reliability prediction. In *Proceedings Annual Reliability and Maintainability Symposium*.

51 Leonard CT, Recht M. (1990). How failure prediction methodology affects electronic equipment design. *Quality and Reliability Engineering International* **6**: 243–249.

52 Wymysłowski A. (2011). Editorial. 2010 EuroSimE international conference on thermal, mechanical and multi-physics simulation and experiments in micro-electronics and micro-systems. *Microelectronics Reliability* **51**: 1024–1025.

53 Kulkarni C, Biswas G, Koutsoukos X. (2010). Physics of failure models for capacitor degradation in DC–DC converters.
https://c3.nasa.gov/dashlink/static/media/publication/2010_MARCON_
DCDCConverter.pdf (accessed January 19, 2018).

54 Eaton DH, Durrant N, Huber SJ, Blish R, Lycoudes N. (2000).
Knowledge-based reliability qualification testing of silicon devices.
International SEMATECH Technology Transfer # 00053958A-XFR.
http://www.sematech.org/docubase/document/3958axfr.pdf (accessed
January 19, 2018).

55 Osterwald CR, McMahon TJ, del Cueto JA, Adelstein J, Pruett J.
(2003). Accelerated stress testing of thin-film modules with SnO_2:F
transparent conductors. Presented at the *National Center for Pho-
tovoltaics and Solar Program Review Meeting Denver,* Colorado.
https://www.nrel.gov/docs/fy03osti/33567.pdf (accessed January 19, 2018).

56 Vassiliou P, Mettas A. (2003). Understanding accelerated life-testing analysis. In *2003 Annual Reliability and Maintainability Symposium.*

57 Mettas A. (2010). Modeling and analysis for multiple stress-type accelerated
life data. In *46th Reliability and Maintainability Symposium.*

58 Dodson B, Schwab H. (2006). *Accelerated Testing: A Practitioner's Guide to
Accelerated and Reliability Testing.* SAE International, Warrendale, PA.

59 Ireson WG, Combs CF Jr, Moss RY. (1996). *Handbook on Reliability Engineering and Management.* McGraw-Hill.

60 Klyatis LM, Klyatis EL. (2006). *Accelerated Quality and Reliability Solutions.*
Elsevier.

61 Klyatis LM, Klyatis EL. (2002). *Successful Accelerated Testing.* Mir Collection, New York.

62 Klyatis LM, Verbitsky D. Accelerated Reliability/Durability Testing as a Key
Factor for Accelerated Development and Improvement of Product/Process
Reliability, Durability, and Maintainability. SAE Paper 2010-01-0203. Detroit.
04/12/2010. (Also in the book SP-2272).

63 Klyatis L. (2009). Specifics of accelerated reliability testing. In *IEEE Workshop Accelerated Stress Testing. Reliability (ASTR 2009)* [CD], October 7–9,
Jersey City.

64 Klyatis L, Vaysman A. (2007/2008). Accurate simulation of human factors
and reliability, maintainability, and supportability solutions. *The Journal of Reliability, Maintainability, Supportability in Systems Engineering*
(Winter).

65 Klyatis L. (2006). Elimination of the basic reasons for inaccurate
RMS predictions. In *A Governmental–Industry Conference "RMS in
A Systems Engineering Environment"*, DAU-West, San Diego, CA,
October 11–12.

66 Klyatis LM. (2012). *Accelerated Reliability and Durability Testing
Technology.* John Wiley & Sons, Inc., Hoboken, NJ.

Exercises

1.1 List three currently used traditional methods of reliability prediction.

1.2 Briefly describe the concept of the basic methods in Example 1.1.

1.3 Why are these methods mostly theoretical?

1.4 Why do most publications in reliability prediction relate to electronics?

1.5 Why are most approaches to reliability prediction of a theoretical nature?

1.6 What is the description of classical test theory?

1.7 What is the basic approach of the Bellcore/Telecordia prediction method?

1.8 What is the basic approach of the MIL-HDBK-17 predictive method?

1.9 What is the basic approach of the ReliaSoft analysis in reliability prediction?

1.10 Describe some of the advantages and disadvantages of empirical methods of reliability prediction.

1.11 Describe the basic concept of the physics-of-failure method of reliability prediction.

1.12 What factors does the Black model for electromigration in electronic products add to reliability prediction modeling?

1.13 What is the basic content of the reliability software modules of the ITEM Toolkit?

1.14 What is the background for the reliability prediction as used by Bellcore for hardware and software?

1.15 What are some of the software and hardware prediction procedures used by Bellcore?

1.16 What procedures are used when employing the Bellcore reliability prediction procedure?

1.17   What problems are encountered with attempts to predict product reliability of mechanical systems using the *Handbook of Reliability Prediction Procedures of Mechanical Equipment*?

1.18   What is the real cause of many recalls?

1.19   What are the basic methods of failure analysis?

1.20   Give a short description of the methods you gave in Exercise 1.19.

1.21   What are the general classes of reliability estimates?

1.22   What are some of the major factors in general model reliability prediction?

1.23   What are the basic strategies of estimating reliability methods?

1.24   Describe some of the advantages and disadvantages of using standards-based reliability prediction.

1.25   Describe some of the advantages and disadvantages of using the physics-of-failure methods.

1.26   What is the reliability factor projected by the life testing method?

1.27   Why are traditional approaches to reliability prediction not successful in industrial practice?

1.28   Why should accurate reliability prediction be so important to a company's management?

1.29   Give some of the reasons traditional methods of reliability prediction in electronics have not been successful?

1.30   Provide a short overview of both qualitative and quantitative aspects of reliability, and their differences.

1.31   Provide a short synopsis of any nine of the articles on reliability prediction from the Reliability and Maintainability Symposium (RAMS) Proceedings that have been discussed in this chapter.

1.32   What are some of the key elements of the Telecordia reliability prediction procedure?

1.33 What is the basic meaning of the Bayesian approach to reliability prediction?

1.34 List several of the basic causes for why traditional solutions using theoretical aspects of reliability prediction are not successful in industrial applications of reliability prediction.

1.35 What is David Nicholls's key conclusion in his published overview of reliability prediction methods?

1.36 Why is fatigue life not an accurate prediction method for successful reliability prediction?

1.37 Describe the basic history of reliability prediction methods.

1.38 Why are most traditional prediction methods related to computer science and manufacture?

1.39 What is the problem with using separate testing aspects to make engineering predictions?

1.40 Why are most presently used approaches failing to produce successful product reliability predictions?

1.41 Describe some of the performance components that interact and affect reliability.

1.42 Why does currently used accelerated stress testing not obtain accurate needed information for successful reliability prediction? Describe some of these causes and their potential effects on prediction.

2

Successful Reliability Prediction for Industry

Lev M. Klyatis

2.1 Introduction

Before we go into the methodology developed by Lev Klyatis, for successful reliability prediction for industry, some background on how this methodology was developed and introduced to industry is helpful. This new direction to successful prediction was created in the Soviet Union, initially for farm machinery, and then expanded into the automotive industry. As success was demonstrated, it expanded into other areas of engineering and industry. Unfortunately, in the Soviet Union much of the work in aerospace, electronics, and defense industries was highly confidential and closed to other professionals, so Dr. Klyatis could not know the availability, or degree of utilization, of his work in these areas. Following successful implementation in the Soviet Union, reliability professionals from American, European, Asian, and other countries would attend Engineering and Quality Congresses, Reliability and Maintainability Symposiums (RAMSs), and Electrical and Electronics (IEEE) and Reliability, Maintainability, and Supportability (RMS) Workshops, where leading professionals in defense, electronics, aerospace, automotive, and other industries and sciences, as well as visiting industrial companies and universities, discussed and learned of these techniques and methodologies.

At professional gatherings such as the annual RAMS, American Society for Quality Annual Quality Congresses, and international conferences, SAE International Annual World Congresses, IEEE symposiums and workshops, RMS symposiums and other workshops, professionals from different areas of industry were exposed to his presentations, and had the opportunity to sit in the room and communicate with each other. These collaboration and discussion helped to advance the science and technology of reliability and its components.

Lev Klyatis recalls visiting the Black & Decker Company as a consultant. In preparation for the visit, professionals involved in different types of testing pre-

Reliability Prediction and Testing Textbook, First Edition. Lev M. Klyatis and Edward L. Anderson.
© 2018 John Wiley & Sons, Inc. Published 2018 by John Wiley & Sons, Inc.

pared questions for Dr. Klyatis. The Director of Quality proposed beginning the meeting with Dr. Klyatis answering the questions. However, Dr. Klyatis suggested an alternate approach. Dr. Klyatis recommended the first step be a review of the testing equipment and technology presently used to obtain a better understanding of their actual processes. Without this step it would be impossible for him to provide the answers to their questions.

This approach was accepted, and together with Black & Decker's engineers and managers he analyzed their existing practices in reliability and testing, including the equipment and the technology they were using. One such example is Dr. Klyatis questioned what was the purpose in doing vibration testing, and what information was obtained from doing the vibration testing. Their response was this testing would provide reliability information for the test subject. He explained, this was not truly a correct answer. Vibration testing is only a part of mechanical testing, which is also only a part of reliability testing. As vibration is only one component of real-world conditions, vibration testing of itself cannot accurately predict product reliability.

A similar situation existed with their test chamber testing. This testing only simulated several of many environmental influences (conditions). As a result of this consulting work, engineers and managers responsible for reliability and testing learned their testing could become more effective. As a result of this analysis, Black & Decker's testing equipment and methodology improved and Dr. Klyatis's consultation was successful.

Similar situations were encountered in dealing with Thermo King Corporation, other industrial companies, and university research centers. As a result of these experiences, Lev Klyatis included these successful approaches and other information he learned from visiting industrial companies and university research centers in his books. Moreover, when Lev Klyatis visited the Mechanical Engineering Department of Rutgers University, he observed single axis (vertical) vibration testing equipment being used to teach students vibration testing. He said to the head if this department that this equipment was used in industrial companies about 100 years ago, and they should be teaching students with advanced vibration equipment with three and six degrees of freedom.

Dr. Klyatis work in consulting and the lessons learned from visiting industrial companies and university research centers helped him to better understand, and analyze in depth, the real-world situation and the state of the art in reliability testing in different areas of industry. This helped further the development of his applications of theory to the development of his successful methodology for prediction of product reliability.

In addition to the value of Dr. Klyatis consulting work, another important aspect in developing this methodology was the role played by seminars where he was an instructor, including his seminars for Ford Motor Company, Lockheed Martin, and other companies. But these experiences are not meant to discount or in any way to minimize the important role of his research of prior publications in the field of simulation and different types of traditional ALT and prediction. From Dr. Klyatis life's work it can be seen that hundreds of references in reliability, testing, simulation, and prediction were studied in the preparation of this textbook.

In fact, through a careful review readers may observe that Dr. Klyatis's published books, papers, and journal articles began with a detailed review and analysis of the current situation. And, as would be expected, not everybody was comfortable with his analysis of the current situation. For example, with previous book publication, John Wiley & Sons selected a person to edit the manuscript. After beginning this work, he decided to stop this editing. He expressed concern that the people and companies analyzed in his book may be offended in being used to demonstrate how their real-world simulation and reliability testing needed improvement. This editor admitted that he did not find anything in the content that was incorrect, but he still had concerns about the critical analysis. While this was a setback, John Wiley & Sons found another, bolder professional who conducted the language editing of the book *Accelerated Reliability and Durability Testing Technology*, which was published in 2012.

With the success of this book, he continued to develop the system of successful prediction of product reliability, especially as related to its strategic aspects.

Dr. Klyatis continued learning how advanced reliability and accelerated reliability/durability testing were being performed through other congresses, symposiums, and tours of the some of the world's most advanced industrial companies. From these experiences, he was surprised to see many world-class companies, such as Boeing, Lockheed Martin, NASA Research Centers, Detroit Diesel, and others, still using older and less advanced methods of reliability testing and reliability prediction for new products and new technologies.

It was observed that the level of ART/ADT was well behind that used by Testmash (Russia), which was described elsewhere [1, 2]. As an example, companies involved in aeronautics and space were conducting mostly flight testing. Laboratory testing primarily consisted of vibration or temperature testing of components (which they were improperly calling environmental testing) and wind tunnel testing for aerodynamic influences.

On several occasions he expressed his concern that their results would be unsuccessful, because their simulation was not duplicating real-world conditions, and they were using the wrong technology in their reliability and durability testing. Some actual examples of this were included in my other published books and will be covered in Chapter 4. It was also observed that some poor reliability prediction work was being performed by some suppliers and universities.

Section 2.2 will demonstrate the step-by-step process to implement an improved method of successful and practical reliability prediction.

2.2 Step-by-Step Solution for Practical Successful Reliability Prediction

It was detailed in Chapter 1 that while there are many publications concerning reliability prediction, many of them are not being used successfully in industry. Section 1.4 considered some of the basic causes for this.

In understanding these causes, one has to understand that:

- In order for a methodology of prediction to work there must be a close combination of reliable sources for calculating the changed reliability parameters for the product's specific models (specimens) during any given time period (warranty period, service life, or others).
- It is very difficult to obtain these necessary parameters of real-world testing. This is because most current field test methods require a very long time to obtain results. For example, the testing time required to obtain reliability parameters that vary during a product's service life.
- Intensive field testing often fails to account for corrosion, deformation, and other degradations that can occur during the long service life of a product. Without these factors the testing cannot provide accurate data to assure successful reliability prediction.
- Generally, accelerated testing in the laboratory and proving grounds does not represent truly accurate simulation of real-world conditions. Therefore, the results of this type of testing will be different from that obtained by field results (see Chapter 3).
- ART/ADT technology does, however, offer this possibility, because it is based on accurate physical simulation of real-world conditions.
- Accurate physical simulation of real-world conditions requires knowledge of a whole gamut of factors that simultaneously and in combination duplicate the complex interaction of real-world influences on the actual product. This must include human factors and risk mitigation.
- The aforementioned items must also relate to subcomponents as well as the complete device, because in the real world they interact. This requires similar testing and reliability prediction by other companies and suppliers.
- Frequently, upper management in organizations is reluctant to embrace using these new scientific approaches. This may be because they are not familiar with them and because they may be concerned that employing them will require serious investments in staffing and equipment. As was written by quality legend Juran, "often they delegate their responsibilities in quality and reliability areas to other people." This problem was addressed in previous publications [1–4].

Figure 2.1 presents a schematic of the step-by-step process for successful practical reliability prediction. By conducting these basic steps and by using the corresponding strategy and methodology, described in this book, one can

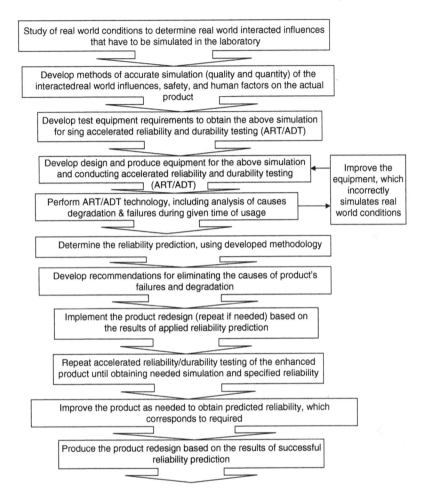

Figure 2.1 Depiction of the step-by-step solution for practical successful reliability prediction.

achieve successful reliability prediction. Some of the details for each step can be found elsewhere [1–3, 5].

As might be expected, the first and absolutely critical step is to study the field conditions and to collect the relevant data that determine the real-world conditions and their interactions that influence the product. Part of this is in the human and safety factors sciences, which also need to be accurately simulated in the laboratory.

The word "interaction" is key, because, in the real world, factors, such as temperature, humidity, pollution, radiation, vibration, and others, do not act separately on the product, but have a combined effect that must be considered.

In order to simulate real-world conditions accurately, these real-world inter-actions must be simulated accurately.

The second step is using the data collected in step 1 to create an accurate simulation (including both quality and quantity) of the interacted real-world influences on the actual product. As used in this step, quality means the simulation must correspond with the criteria of accurate usage. This includes divergences from the theoretical or assumed use as was established by the researchers or designers. A detailed description of these criteria can be found in this chapter. Quantity means an accurate number of relevant input influences, which are needed to simulate real-world conditions in the laboratory. As previously indicated, this depends on the number of field input influences.

2.3 Successful Reliability Prediction Strategy

The common scheme of reliability prediction strategy for industry, if one wishes to obtain successful prediction results, can be seen in Figure 2.2.

Depiction of the five common steps for successful reliability prediction (Figure 2.3) consists of:

1. accurate physical simulation of the field conditions;
2. ART/ADT technology;
3. methodology of reliability prediction;
4. successful reliability prediction;
5. successful reliability prevention and development.

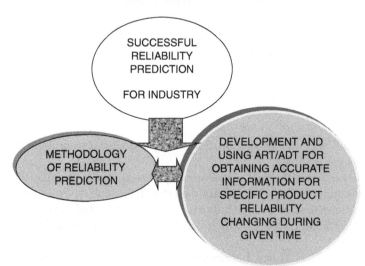

Figure 2.2 Common scheme of successful reliability prediction strategy.

Figure 2.3 Five common steps for successful reliability prediction.

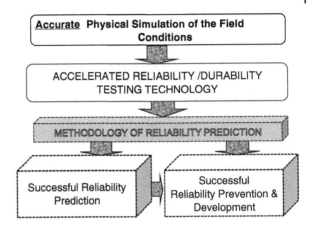

Lev Klyatis created his ideas of successful reliability prediction while working in the USSR, where he developed some of the components, methodology, and tools for this new scientific engineering direction. Then he continued his work in this field after moving to the USA, and which he continues developing to this day, improving the direction, strategy, and continuing the expansion of its implementation.

As was shown in Figure 2.2, this strategy includes two basic components, the details of which were published elsewhere [1–3]. The first component to successful prediction methodology is detailed and described in Section 2.5. The second component was described in detail in *Accelerated Reliability and Durability Testing Technology* [1]. The basics of this second component will also be described in Chapter 3.

Figure 2.4 demonstrates the interaction of the three basic groups of real-world conditions necessary for successful reliability prediction.

2.4   The Role of Accurate Definitions in Successful Reliability Prediction: Basic Definitions

Accurate definitions of terminology is a key factor in conducting successful reliability prediction. In his previous books this author included some examples of how misunderstanding of definitions can lead to inaccurate prediction [1, 3]. Earlier in this chapter, an example of this author's experience with Black & Decker was presented. It was also described [3] how professionals from the Nissan Technical Center considered vibration testing as reliability testing, while not accounting for other significant field input influences. Because these definitions were not properly understood, the results of the testing were not accurate predictors of product performance.

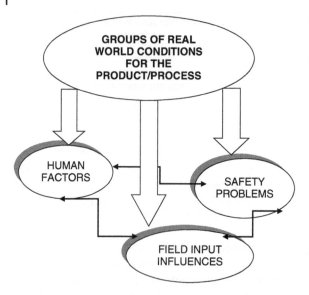

Figure 2.4 Interacted groups of real world conditions for the product/process.

Box 2.1 gives the correct basic definitions that reliability professionals need to know for accurate reliability testing and prediction. Additional definitions can be found in Chapter 4, which includes the draft standard which has been approved at SAE G-11 Reliability Committee meetings.

Box 2.1 Correct basic definitions that reliability professionals need to know	
Accelerated testing	Testing in which the deterioration of the test subject is accelerated.
ART (or ADT or durability testing)	Testing in which: a) The physics (or chemistry) of the degradation mechanism (or failure mechanism) is similar to this mechanism in the real world using given criteria. b) The measurement of reliability and durability indicators (time to failures, its degradation, service life, etc.) has a high correlation with real-world use (corresponding given criteria).
Note 1	ART, or ADT, or durability testing provides useful information for accurate prediction of product reliability and durability, because they are based on accurate simulation of real-world conditions.
Note 2	This testing is identical to reliability testing if reliability testing is used for accurate reliability and durability prediction during service life, warranty period, or other specified time of utilization.

Note 3	ART/ADT is related to the degree of stress used in the process. A higher level of stress will result in higher acceleration coefficient (ratio of the time to failures of the product in the field compared with the time to failures during ART), while a lower level of stress will result in a lower correlation between field results and ART, thus yielding a less accurate prediction.
Note 4	ART and ADT (durability testing) consists of:
	A combination of a complex and comprehensive laboratory testing combined with periodical field testing.
	The laboratory testing must be designed to provide a simultaneous combination of the entire complex of interacted multi-environmental testing, mechanical testing, electrical testing, and other types of real-world testing.
	The periodic field testing takes into account the factors which cannot be accurately simulated in the laboratory, such as the stability of the product's technological process, how the operator's reliability has an influence on the test subject's reliability and durability, changing cost during usage time, and others.
	Accurate simulation of field conditions requires full understanding of the simulation of the field input influences integrated with safety and human factors.
Note 5	ART and ADT (or durability testing) have the same desired outcome—the accurate simulation of the product's performance in a field situation.
	The primary difference is in the metrics used for these types of testing and the length of testing. For reliability, the outcome is generally expressed in the MTTFs, time between failures, and similar parameters; while for durability it is a measure of the product's uptime or expected in-service time.
Note 6	ART can be performed for different lengths of time, such as warranty period, a regulatory mandated period, 1 year, 2 years, service life, and others.
Accurate prediction	This is possible if one has:
	a) prediction methodology to incorporate all active field influences and interactions;
	b) accurate initial information (from ART/ADT) for calculation of changing predicted parameters of each product model during given time.
Accurate simulation of the field input influences	If all field influences act simultaneously and in mutual combination, and accurately simulated with divergence no more than given limit.

(Continued)

Box 2.1 (Continued)	
Accurate system of reliability prediction	The system of prediction is accurate if, and only if, the simulation of field conditions is accurate and ART is possible.
	It consists of two basic components: (a) methodology and (b) a source for obtaining accurate initial information for calculation of reliability changes during a given time.
Field test	A test administered and used to check the adequacy of testing procedures in the actual normal service, generally including test administration, test responding, test scoring, and test reporting.
Multi-environmental complex of field input influences	Includes temperature, humidity, pollution, radiation, wind, snow, fluctuation, and rain, and other effecting factors.
	Input influences often combine to form an effect very different than if tested separately and independently.
	For example, chemical pollution and mechanical pollution may combine in the pollution variable and produce greater product degradation than would be evidenced through independent testing. These seemingly interdependent factors are actually interconnected, and interact simultaneously and in combination with each other.
Accelerated corrosion testing	Is testing with simultaneous simulation interactions, including:
	• chemical pollution • mechanical pollution • moisture • temperature • vibration • deformation • friction
Vibration testing	Is testing with all vibration factors applied simultaneously.
	For example, for a road vehicle the simulation interactions would include:
	• features of the road, including profile, types of the road, road density; • design and quality of test subject; • speed and direction of the wind; • speed of test subject; • design and quality of the wheels and suspension.

2.5   Successful Reliability Prediction Methodology

The basic methodology for successful reliability prediction includes:

- common criteria for successful prediction of product reliability;
- methodology for selecting representative input regions for accurate simulation of real-world conditions;
- aspects of successful prediction of product reliability, using manufacturing technology factors and usage conditions;
- building an appropriate testing model for reliability prediction;
- system reliability prediction from testing results of the components.

A depiction of the common scheme of this methodology can be seen in Figure 2.5.

Klyatis and Walls [6] provide further information on the methodology for selecting representative input regions for accurate simulation of real-world conditions.

2.5.1   Criteria of Successful Reliability Prediction Using Results of Accelerated Reliability Testing

The results of ART are frequently used as a source for obtaining initial information needed to predict the reliability of machinery in field conditions. But for this to be accurate one must be sure that the prediction is correct (if possible, with a given accuracy). The following solution is useful in achieving this goal.

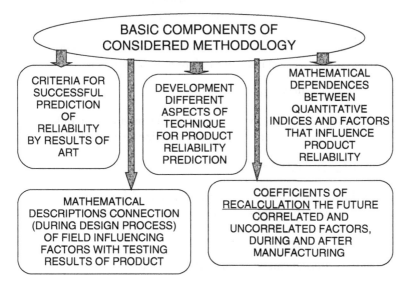

Figure 2.5 Common scheme of methodology for product's reliability successful prediction.

The problem is formulated as follows: there is the system [results of use of the current product in the field] and its model [results of ART/ADT of the same product]. The quality of the system can be estimated by the random value φ using the known or unknown law of distribution $F_S(x)$. The quality of the model can be estimated by the random value ϕ using the unknown law of distribution F_M. The model of the system will be satisfactory if the measure of divergence between F_S and F_M is less than a given limit Δ_g.

After testing the model, one obtains the random variables $\varphi_1: \varphi_1^{(1)}, \dots, \varphi_1^{(n)}$. If one knows $F_S(x)$, by means of $\varphi_1^{(1)}, \dots, \varphi_1^{(n)}$, one needs to check the null hypothesis H_0, which means that the measure of divergence between $F_S(x)$ and $F_M(x)$ is less than Δ_g. If $F_S(x)$ is unknown, it is necessary also to provide testing of the system. As results of this testing, one obtains realizations of random variables $\varphi: \varphi^{(1)}, \dots, \varphi^{(m)}$. For the aforementioned two samplings, it is necessary to check the null hypothesis H_0 that the measure of divergence between $F_S(x)$ and $F_M(x)$ is less than the given limit Δ_g. If the null hypothesis H_0 is rejected, the model needs updating; that is, to look at more accurate ways of simulating the basic mechanism of field conditions use for ART.

The measure of divergence between $F_S(x)$ and $F_M(x)$ is functional as estimated by a multifunctional distribution and depends on a competitive (alternate) hypothesis. The practical use of the criteria obtained depends on the type and form of this functional. To obtain exact distributions of statistics on the condition that the null hypothesis H_0 is correct is a complicated and unsolvable problem in the theory of probability. Therefore, here, the upper limits are shown for the statistics and their distributions studied, so the level of values will be increased; that is, explicit discrepancies can be detected. Let us consider the situation when $F_S(x)$ is known.

First, we will take as the measure of divergence between the functions of distribution $F_S(x)$ and $F_M(x)$ the maximum of modulus difference:

$$\Delta[F_M(x),\ F_S(x)] = \max_{(x)<\infty} |F_M(x) - F_S(x)||$$

We understand that H_0 is the hypothesis that the modulus of difference between $F_M(x)$ and $F_S(x)$ is no more than the acceptable level Δ_g; that is:

$$\overline{H}_0 : \max_{(x)<\infty}[F_M(x) - F_S(x)] \le \Delta_g$$

where $F_M(x)$ is the model (testing conditions) function of distribution.

Against H_0 one checks the competitive hypothesis:

$$\overline{H}_1 : \max|F_M(x) - F_S(x)| > \Delta_g$$

The statistic of the criterion can be given by the formula

$$\overline{D}_n = \max_{(x)<\infty} |F_M(x) - F_S(x)|$$

Practically, it can be calculated by the following formula:

$$\overline{D}_n = \max_{1 \le m \le n} \left\{ \max\left[\frac{m}{n} - F(\eta_m)\right], \max\left[F(\eta_m) - \frac{m}{n}\right] \right\}$$

It is very complicated to find the distribution of this statistic directly [7]. The $D_n \to \Delta_g$ as $n \to \infty$. Therefore, it is necessary to look for the distribution of a random value $\sqrt{n(D_n - \Delta_g)}$.

Let us give the upper estimation which can be useful for practical solution of this problem:

$$\overline{D}_n = \max_{(x)<\infty}[F_n(x) - F_S(x)]$$

$$= \max[F_n(x) - F_M(x) + F_M(x) - F_S(x)] \le \max_{(x)<\infty}\{|F_n(x)|\}$$

$$\{|F_M(x)| + |F_S - F_M(x)|\} \le \max_{(x)<\infty}|F_n(x) - F_M(x)| + \max_{(x)<\infty}|F_M(x) - F_S(x)|$$

If hypothesis H_0 is true, then

$$\max_{(x)<\infty}|F_M(x) - F_S(x)| \le \Delta_g$$

Therefore:

$$\overline{D}_n \le \max_{(x)<\infty}|F_n(x) - F_M(x)| + \Delta_g$$

or

$$\sqrt{n(\overline{D}_n - \Delta_g)} \le \sqrt{n}\max_{(x)<\infty}|F_n(x) - F_M(x)| \qquad (2.1)$$

Here, if $F(x)$ is the probability of work without failure, then n is the number of failures.

Let us denote $\max_{(x)<\infty}|F_n(x) - F_M(x)|$ as D_n. This random value $\sqrt{nD_n}$ limited by $n \to \infty$ follows Kolmogorov's law [8]. Therefore:

$$P\left\{\sqrt{n(\overline{D}_n - \Delta_g)} < x\right\} \ge K(x)$$

or

$$P\left\{\sqrt{n(\overline{D}_n - \Delta_g)} \ge x\right\} < 1 - K(x)$$

where $K(x)$ is the function of the Kolmogorov distribution.

As a result of research, the following is the correct way to use the Kolmogorov criterion. First, one calculates the number $\sqrt{n(\overline{D}_n - \Delta_g)} = \lambda_0$. Then:

$$P\left\{\sqrt{n(\overline{D}_n - \Delta_g)} \ge \lambda_0\right\} < 1 - K(\lambda_0)$$

If the difference $1 - K(\lambda_0)$ is small, then the probability $P\left\{\sqrt{n(D_n - \Delta_g)} \geq \lambda_0\right\}$ is also small. This means that an improbable event has occurred, and the divergence between $F_n(x)$ and $F_S(x)$ can be considered as a substantial, rather than a random, character of the values studied and Δ_g. Therefore, the conclusion is

$$\max_{(x)<\infty} |F_S(x) - F_M(x)| > \Delta_g$$

If the level of value of this criterion is higher than $1 - K(\lambda_0)$, hypothesis H_0 is rejected. If $1 - K(\lambda_0)$ is large, it does not exactly confirm the hypothesis, but by a small Δ_g we can practically consider that the testing results do not contradict the hypothesis.

Let us take now as a measure of divergence between $F_S(x)$ and $F_M(x)$ the maximum difference between $F_S(x)$ and $F_M(x)$ by the Smirnov criterion [9]; that is:

$$\Delta[F_M(x), F_S(x)] = \max_{(x)<\infty}[F_M(x) - F_S(x)]$$

In this case the hypothesis H_0 becomes

$$H_0 : \max_{(x)<\infty}[F_n(x) - F_S(x)] \leq \Delta_g$$

The statistic of the criterion is [5]

$$\overline{D}_n^+ = \max\left[\frac{m}{n} - F(\eta_m)\right]$$

By analogy with the previous solution, the upper value is

$$\overline{D}_n^+ = \max_{(x)<\infty}|F_n(x) - F_M(x)| + \Delta_g$$

or

$$\sqrt{n\left(\overline{D}_n^+ - \Delta_g\right)} \leq n\max_{(x)<\infty}|F_n(x) - F_M(x)| \times \sqrt{nD_n^+}$$

The random value $\sqrt{nD_n^+}$ in the limit has a Smirnov distribution; therefore:

$$P\left\{\sqrt{n(\overline{D}_n^+ - \Delta_g)} < x\right\} \geq 1 - e^{-2x^2}$$

or

$$P\{n\left(\overline{D}_n^+ - \Delta_g\right) > e^{-2x^2}$$

As a result, the following rule for use of the criterion was obtained. First, one calculates the value $\sqrt{n(D_n^+ - \Delta_g)} = \lambda_0$. Then

$$\left\{\sqrt{n\left(\overline{D}_n^+ - \Delta_g\right)} \geq \lambda\right\} < e^{-2\lambda_0^2}$$

If $e^{-2\lambda_0^2}$ is small, therefore the probability $P\left\{\sqrt{n(\overline{D}_n^+ - \Delta_g)}\right\} \geq \lambda_0$ is also small, and if we describe it as analogous with the above, then

$$\max[F_M(x) - F_S(x)] > \Delta_g$$

In consideration of the alternate hypothesis H_1^-, everything will be analogous to hypothesis H_1^+, because if the minuend and subtrahend exchange places, that will not change the final result.

Second, let us consider checking the hypothesis H_0 by using the alternate hypothesis $H_1[\varphi(F)]$ with only the weight function

$$\varphi(F) = \frac{1}{F_S(x)} \quad \text{if } F_S(x) \geq a$$
$$0 \quad \text{if } F_S(x) < a$$

Let us take as a measure of divergence between $F_S(x)$ and $F_M(x)$

$$\Delta[F_S(x), F_M(x)] = \max_{F_S(x) \geq a} \frac{|F_M(x) - F_S(x)|}{F_S(x)}$$

The statistics of the criterion can be expressed by

$$R_n(a, 1) = \max_{F_S(x) \geq a} \frac{F_n(x) - F_S(x)}{F_S(x)}$$

For this practical calculation the following formula can be used:

$$\overline{R}_n(a, 1) = \max\left\{ \max_{F(\eta_M) \geq a} \frac{\frac{m}{n} - F(\eta_M)}{F(\eta_M)}, \max_{F(\eta_M) \geq a} \frac{F(\eta_M) - \frac{m}{n}}{F(\eta_M)} \right\}$$

As stated earlier, the upper value for this statistic was found to be

$$\overline{R}_n(a, 1) = \max_{F_S(x) \geq a} \frac{|F_n(x) - F_S(x)|}{F_S(x)} \leq \max_{F_M(x)} \frac{F_n(x) - F_M(x)}{F_M(x)}$$

$$\times \max_{F_S(x) \geq a} \frac{F_M(x)}{F_S(x)} + \Delta_g \leq \overline{R}_n(a, 1)\frac{1}{a} + \Delta_g \qquad (2.2)$$

Hypotheses H_0 and H_1 then become

$$H_0 : \max_{F_S(x) \geq a} \frac{|F_n(x) - F_S(x)|}{F_S(x)} \leq \Delta_g$$

$$H_1 : \max_{F_M(x) > a} \frac{|F_n(x) - F_S(x)|}{F_S(x)} > \Delta_g$$

We obtain Equation 2.2, because

$$\max_{F_M(x) > a} \frac{|F_n(x) - F_M(x)|}{F_M(x)}$$

is a statistic of $R_n(a, 1)$.

Therefore, the random value $\sqrt{n[R_n(a, 1) - \Delta_g]}$, limited by $n \to \infty$, follows the law of distribution [7]:

$$L(x) = \frac{4}{\pi} \sum_{k=0}^{\infty} \frac{(-1)^k}{2^{k+1}} e - \frac{(2k+1)^2 \pi^2}{8x^2}$$

From here

$$P\{\sqrt{na\left[\overline{R}_n(a, 1) - \Delta_g\right]} < x$$

or

$$P\left\{\sqrt{n\left[\overline{R}_n(a, 1) - \Delta_g\right]} \geq x\right\} < 1 - 4(\lambda_0)$$

So, we obtained the Klyatis criteria that are modifications of the Kolmogorov and Smirnov criteria.

There is the following rule for using Klyatis's criteria:

(A) One calculates the actual number $\sqrt{na[R_n(a, 1) - \Delta_g]} = \lambda_0$.

(B) In that case:

$$P\left\{\sqrt{na[R_n(a, 1) - \Delta_g]} \geq \lambda_0\right\} < 1 - L(\lambda_0)$$

(C) If $1 - L(\lambda_0)$ is small, then the probability $P\{\sqrt{na[R_n(a, 1) - \Delta_g]} \geq \lambda_0$ will also be small. This means that the difference between $F_n(x)$ and $F_S(x)$ is significant.

Then we will take the maximum difference as a measure of divergence:

$$\Delta[F_M(x), F_S(x)] = \max_{F_S(x) \geq a} \frac{F_M(x) - F(x)}{F_S(x)}$$

This problem can also be solved by the method analogous to the previous solution. The rule of use of this criterion is as follows. One calculates the actual number

$$\sqrt{na[R_n^+(a, 1) - \Delta_g]} = \lambda_0$$

Then

$$P\left\{\sqrt{na[R_n^+(a, 1) - \Delta_g]} \geq \lambda_0\right\} < 2\left[1 - \Phi\left(\frac{\lambda_0 \times \sqrt{a}}{\sqrt{1 - a}}\right)\right]$$

If $2\{1 - \Phi[(\lambda_0 \times \sqrt{a})/\sqrt{1 - a}]\}$ is small, it means that hypothesis H_0 is rejected, and then by analogy with previous solutions the same applies to the competitive hypothesis H_1.

All are analogous with weight function

$$\psi[F_S(x)] = \begin{cases} 1 & \text{when } F_S(x) \le a \\ 1 - F_S(x) & \text{when } F_S(x) > a \end{cases}$$

Let us now consider the variant when $F_S(x)$ is unknown. In this case one provides an ART of the prediction subject. As a result, one obtains realizations of the random value φ of system reliability, $\varphi^{(1)}, \ldots, \varphi^{(m)}$, and can use these realizations to build functions of the distribution $F_m(x)$. By means of the functions of distribution $F_n(x)$ and $F_m(x)$, it is necessary to establish whether the random value studied relates to one class or not; that is, will the divergence between the actual functions of distribution $F_n(x)$ and $F_S(x)$, by a certain measure, be less or more than the given tolerance Δ_g.

Let us take the measure of divergence in the Smirnov criterion as a measure of divergence between functions of distribution [9]:

$$\Delta[F_M(x), F_S(x)] = \max_{(x) < \infty} [F_M(x) - F_S(x)]$$

In this case the null hypothesis H_0 looks like

$$\widetilde{H_0} : \max_{(x) < \infty} [F_n(x) - F_m(x)] \le \Delta_g$$

The alternative hypothesis H_1^+ looks like

$$H_1 : \max[F_n(x) - F_m(x)] > \Delta_g$$

The statistic of the criterion can be expressed by the following formula:

$$\tilde{D}_{m,n} = \max_{(x) < \infty} [F_n(x) - F_m(x)]$$

Its upper estimation

$$\tilde{D}_{m,n}^+ = \max_{(x) < \infty} \{F_n(x) - F_S(x) - F_m(x) + F_M(x) + F_S(x) - F_M(x)\}$$

$$\le \max[F_n(x) - F_S(x)] + \max_{(x) < \infty} [F_M(x) - F_m(x)] + \max_{(x) < \infty} [F_S(x) - F_M(x)]$$

If hypothesis H_0 is true, then

$$\overline{D}_{m,n}^+ \le D_n^+ + D_m^- + \Delta_g$$

where

$$\overline{D}_n^+ = \max_{(x) < \infty} [F_n(x) - F_S(x)] \quad \text{and} \quad \overline{D}_m^+ = \max_{(x) < \infty} [F_M(x) - F_m(x)]$$

Statistics $\overline{D}_m^+$ and $\overline{D}_n^+$, as was shown earlier, have equal distributions. Therefore:

$$\overline{D}_{m,n}^+ - \Delta_g \le D_n^+ + D_m^n$$

Therefore:

$$\sqrt{\frac{mn}{m+n}}(D^+_{m,n} - \Delta_g) \le \sqrt{\frac{mn}{m+n}}D^+_n + \sqrt{\frac{mn}{m+n}}D^+_m$$

Let n and m approach infinity, so that $m/n \to k$. Then

$$\lim_{n \to \infty} \sqrt{\frac{mn}{m+n}}(D^+_{m,n} - \Delta_g) \le \sqrt{\frac{1}{1+k}} \lim_{n \to \infty} \sqrt{nD^+_n} + \sqrt{\frac{k}{k+1}} \lim_{n \to \infty} D^+_m$$

Let us denote the random variable $\lim \sqrt{nD^+_n}$ through V_2. V has a Smirnov function of distribution $F_V(x) = 1 - e^{-2x^2}$. Then from the assumption that m and n are large enough, this solution has already been published by Klyatis [1, 2]. As a result, we obtain the following rule of criterion use.

To calculate the actual number

$$\sqrt{\frac{mn}{m+n}}(\overline{D}^+_{m,n} - \Delta_g) = \lambda_0 \tag{2.3}$$

where n is number of failures during ART/ADT and m is the number of failures in the field.

Therefore:

$$P\left\{\sqrt{\frac{mn}{m+n}}(\overline{D}^+_{m,n} - \Delta_g) \ge \lambda_0\right\} \le 1 - F_\xi(\lambda_0)$$

If $1 - F_\xi(\lambda_0)$ is small, the hypothesis H_0 is rejected by analogy with the previous calculations. In this case $\xi = V$ if $k = 0$ or $k = \infty$.

If we take as a measure of divergence between distributions of functions a functional

$$\Delta[F_S(x), F_M(x)] = \max_{x < \infty} |F_M(x) - F_S(x)|$$

and make the actions analogous to previously, then we obtain the rule of use for the criterion in the following form. First, we calculate the following number:

$$\sqrt{\frac{mn}{m+n}}(\overline{D}^+_{m,n} - \Delta_g) = \lambda_0$$

Then

$$P\left\{\sqrt{\frac{mn}{m+n}}(\overline{D}^+_{m,n} - \Delta_g) \ge \lambda_0\right\} \le 1 - F_x(\lambda_0)$$

If $1 - F_x(\lambda_0)$ is small, it means that the hypothesis H_0 is rejected, and then is analogous to previous actions.

For finding the distribution of random variable k, one can use special dependences, where

$$a = \frac{k_1}{\sqrt{k+H}}; \qquad b = \frac{\sqrt{k_1}}{\sqrt{k_1 + 1}}$$

in the function of Kolmogorov's distribution

$$F_\xi(k) = \sum_{k=-\infty}^{\infty} (-1)^k e^{-2k2x^2}$$

In conclusion:

1. The engineering version of the solution obtained is that the upper estimation of the statistical criteria of correspondence, for some measures between the functions of distribution of studied reliability characteristics were created in ART conditions and in field conditions. This can be useful for practical reliability prediction for industry as well as for solving other engineering problems (accelerated reliability development and improvement, etc.).

2. The mathematical version of the solution obtained is that approximate Klyatis criteria as modifications of Smirnov's and Kolmogorov's criteria by divergence ($\Delta_g < 0$) were obtained for comparison of two empirical functions of distribution by measurement of the Smirnov divergence

$$\Delta[F_S(x), F_M(x)] = \max_{(x)<\infty} [F_M(x) - F_S(x)]$$

and the Kolmogorov divergence

$$\Delta[F_S(x), F_M(x)] = \max_{(x)<\infty} |F_S(x) - F_M(x)|$$

In the Smirnov criterion by zero hypothesis, we have

$$\max_{(x)<\infty} [F_M(x) - \widetilde{F_m}(x)] < \Delta_g$$

By the competitive hypothesis, we have

$$\max_{(x)<\infty} [F_M(x) - \widetilde{F_m}(x)] > \Delta_g$$

If $\Delta_g = 0$, we have the Smirnov criterion. An analogous situation applies for the Kolmogorov criterion. The difference between the two versions is that in the measure using the Klyatis modification of the Smirnov criterion one takes into account only those regions (the oscillograms of loadings, etc.) where $F_S(x) > F_M(x)$ and one looks for maximum differences only for them.

In measuring with the Klyatis modification of Kolmogorov's criterion one takes into account the maximum differences for all regions by modulus. The consideration of both criteria makes sense, because Smirnov's criterion is easier to calculate, but does not give the full picture of divergences between $F_S(x)$ and $F_M(x)$; Kolmogorov's criterion gives a fuller picture of the above divergence, but is more complicated in calculation.

Therefore, the choice of the better criterion for a specific situation must be decided according to the specific conditions of the problem to be solved.

Let us show the solution obtained by a practical example. In the field, the details for $m = 102$ failures of car transmissions were obtained. After ART/ADT, 95 failures were obtained: $n = 95$, $\Delta_g = 0.02$.

In the field one builds the empirical function of distribution of the time to failures $F_m(x)$ by the intervals between failures, and one builds by intervals between failures during ART/ADT of the function of distribution time to failures $F_M(x)$. As we can see, this is the last variant to be considered

If we align the graph $F_M(x)$ (Figure 2.1) and the graph $\widetilde{F}_m(x)$, we will find the maximum difference between $F_M(x)$ and $F_m(x)$. For this goal we can draw the graph $F_m(x)$ on transparent paper and it is simple to find the maximum difference $D_{m,n}^{+} = 0.1$. In correspondence with Ventcel [10], we have $\lambda_0 = 0.98$:

$$k = \frac{m}{n} \approx 1$$

and therefore

$$F_x(x) = 1 - e^{-2x^2} \left[1 + x\sqrt{2\pi}\Phi(x) \right]$$

After substitution of $\lambda_0 = 0.98$, we obtain $F_x(0.98) = 0.6$. Therefore, $1 - F_x(0.98) \approx 0.4$. So, $1 - F_x(0.98)$ is not small and the hypothesis H_0 can be accepted. Therefore, the divergence between actual functions of distribution of time to failures of the aforementioned test subject (e.g., car transmission) details for the car tested in field conditions and in ART/ADT conditions by Smirnov's measure is within the given limit $\Delta_g = 0.02$ (Figure 2.6). This gives

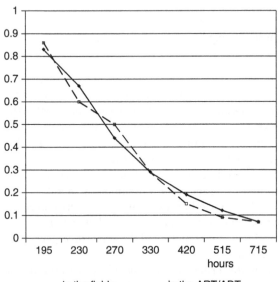

Figure 2.6 Evaluation of the correspondence between functions of distribution of the time to failure of a car trailer's transmission details in the field and in the ART/ADT conditions.

− − − in the field ; ———— in the ART/ADT

the possibility for successful prediction of reliability of the car transmissions using the results of this testing.

2.5.2 Development of Techniques for Product Reliability Prediction Using Accelerated Reliability Testing Results

This section will address the problem of solving the successful prediction of product reliability taking into account the effects of reliability with complex input factors.

The typical practice in engineering is to test a small sample number (from five to ten test specimens of each component) with two to five possible failures deemed as acceptable. Usually, it is assumed that the failures of the system components are statistically independent.

This proposed approach is very flexible and useful for many different types of products, including electronic, electromechanical, mechanical, and others.

2.5.2.1 Basic Concepts of Reliability Prediction

As was mentioned earlier in this book, in order for reliability prediction to be useful, it must be based on the appropriate methodology, techniques, and equipment to assure accurate initial information for the prediction.

The methodology was partially considered earlier.

ART/ADT can give this information if one follows the successive step-by-step technology described in this book.

The basic concept of successful reliability prediction consists of the following basic steps:

1. Building an accurate model of real-time performance.
2. Using the model for testing the product and studying the degradation mechanism over time and comparing model degradation with the real-life degradation mechanism of the product. If the degradation mechanism differs by more than a fixed defined limit, one must improve the model's real-time performance.
3. Making real-time performance forecasts for reliability prediction using these testing results as initial information.

Each step can be performed in different ways, but reliability can only the predicted accurately if the researchers and engineers use this concept.

Step 1 can only be performed if one understands that real-life reliability of the product depends on a combination of different interacted input influences, such as is shown in Chapter 3. The simulation of input influences must be as complicated as they are in real-life conditions. For example, for a mobile product one needs to use multi-axis vibration in combination with multi-environmental and other factor testing.

In order to solve step 2 one has to understand the degradation mechanism of the product and the parameters affecting this mechanism. The causes of the product's degradation mechanism are included in the data, for example the data must include, the electrical, mechanical, chemical, thermal, and radiation effects. And some of the parameters of the mechanical degradation include deformation, crack, wear, creep, and so on. In real life, different processes of degradation may be acting simultaneously and in combination.

Therefore, ART must also include the simultaneous combination of different types of testing (environmental, electrical, vibration, etc.), with the assumption that the failures are statistically related to these combined factors. The degradation mechanism of the product during ART/ADT must be closely similar to the mechanism that occurs in real life.

In order to solve step 3 of this reliability prediction technique, all pertinent aspects must be considered, including both manufacturing and field conditions.

2.5.2.2 Prediction of the Reliability Function without Finding the Accurate Analytical or Graphical Form of the Failures' Distribution Law

The problem was solved for two types of conditions: (a) prediction consisting of point expressions of reliability function of the system elements; (b) prediction of the reliability functions of the system with predetermined accuracy and confidence area [11].

Problem (a) can be solved with grapho-analytical methods on the basis of failure hazard or frequency if we have the graph $f(t)$ of the empirical frequency of failures; then, guided by the failure frequency graph, one can discover the reliability function:

$$p(t) = \int_0^t f(t) \ dt = 1 - S_f \tag{2.4}$$

where $\int_0^t f(t) \ dt = S_f$ is the area under the curve $f(t)$ that was obtained as a result of ART.

The reliability function of a system which consists of different components (details) is:

$$P(t) = \prod P_j(t) = \prod (1 - S_{fj}) \tag{2.5}$$

For example, as a result accelerated testing of belts $t = 250$, the area is $S_f = 1.12$, and probability $P(t) = 0.82$.

In variant (b) one needs to calculate the accumulated frequency function and the values of the confidence coefficient found in the equations:

$$Y(x) = \sum_{m}^{n} C_n^m p^m (1 - P)^{n-m} \tag{2.6}$$

$$Y(x) = \sum_{m=0}^{k} C_n^m p^m (1 - P)^{n-m} \tag{2.7}$$

and evaluate the curves that are limited to the upper and lower confidence areas. In Equations 2.6 and 2.7, $C_n^m p^m (1 - P)^{n-m}$ is the probability that, based on an event, there will be in n independent experiments m times. The values of $\overline{Y}$ and $\underline{Y}$ are found in the tables of the books on the theory of probability if the confidence coefficient is $\lambda = 0.95$ or $\lambda = 0.99$.

2.5.2.3 Prediction Using Mathematical Models Without Indication of the Dependence Between Product Reliability and Different Factors of Manufacturing and Field Usage

These factors (influences) should be evaluated as the results of ART/ADT. The solution of this problem is possible using mathematical models which best describe the dependence between the reliability and the series of factors shown above:

$$z_i^{(\tau)} = f_i^{(\tau)}(v_1; v_2; \dots ; v_\theta) \qquad \tau = 1, 2, \dots, n \tag{2.8}$$

where θ is the number of all factors, n is the number of reliability indexes which most completely characterize the ith model of the product, $f_i^{(\tau)}(v)$ is the function that gives the possibility of finding the optimum value of these functions indexes, by changing the level of factors influencing the products' reliability. A large amount of statistical data is necessary to build this function, which can be obtained by experiments with fixed values of factors for estimation of reliability indexes.

This is a difficult problem that requires a large expenditure of resources for its solution. Prediction of reliability indexes for a product which will be manufactured in the future is more readily obtained by results of ART/ADT of its early specimens. In this case, one part of the input influences (mechanical, environmental, etc.) can be used to obtain results of product ART/ADT. Another part (influences of manufacturing specifics, conditions of operator specifics, etc.) must be taken into account when studying the results of mathematical modeling.

In this case, the connection between evaluation of $Z_{ij}^{(\tau)}$ of the τth quantified reliability index of the ith model of product in the jth conduction of use, which needs of prediction, and the mean value of this index $Z_i^{(\tau)}$, which can be obtained as a result of accelerated testing of μ specimens of the product, is described with functional dependence

$$Z_{ij}^{(\tau)} = G^{(\tau)}(Z_i^{(\tau)}; U_i(t); a_j^{(\tau)}; a_{ij}^{(\tau)}) \qquad \tau = 1, 2, \dots, n \tag{2.9}$$

where $U_i(t) = F_{mi}(t)$ and $F_{ij}(t)$ are the most important lack of correlated common factors of manufacture and in the field; $a_i^{(\tau)}$ and $a_{ij}^{(\tau)}$ are unknown parameters of the mathematical model in Equation 2.9 which are characterized by the group of manufacture and field factors

$$F_{mi}(t) = F_{m1}(t); F_{m2}(t); \dots ; F_{mn}(t)$$
$$F_{ij}(t) = F_{i1}(t); F_{i2}(t); \dots ; F_{in}(t)$$

$m = 5$ and $i = 6$ are the numbers of the most important lack of correlated common factors of manufacturing and the field.

The mean value $Z_i^{(\tau)}$, which is a result of ART/ADT of μ specimens (usually not more than two or three specimens), can be evaluated as

$$Z_i^{(\tau)} = \mu^{-1} \sum_{k=1}^{m} Z_{ik}^{(\tau)} \tag{2.10}$$

The results of the aforementioned testing are independent of factors that cannot be simulated in the laboratory. Therefore, in the model in Equation 2.9 the variable can be divided into

$$Z_i^{(\tau)} = K_{ij}^{(\tau)}(F_{ni}(t); F_{fi}(t); a_j^{(\tau)}; a_{ij}^{(\tau)})Z_i^{(\tau)} \tag{2.11}$$

where $K_{ij}^{(\tau)}$ are the recalculated coefficients of the quantitative values' quantitative index of future product reliability concerning the indexes of means indexes which have been obtained by ART of specimens.

These coefficients depend on manufacturing and field conditions, which themselves are time dependent and contain unknown parameter values. The values of these coefficients are different for different products and different indexes of reliability. This is because the levels of the most important factors of manufacturing by different companies and in different field conditions are not the same. The values of recounting coefficients may be more or less than one for different reliability indexes. Therefore, the coefficients are functionals:

$$K_{ij}^{(\tau)} = F^{(\tau)}\{f[F_{ni}(t); a_i^{(\tau)}; a_{ij}^{(\tau)}]\} \tag{2.12}$$

where $F[F_{ni}(t); F_{fi}(t); a_i^{(\tau)}; a_{ij}^{(\tau)}]$ is the function which evaluates the level of influence on the product reliability of the basic lack of correlated generalized factors of the field and manufacturing conditions. The most important factors are characterized by ponderable levels P and Q and actual levels X_i and Y_i.

Let us consider the function of the impact of the statistical problem of product reliability prediction if we take into account the weak dependence of these factors on time:

$$K_{ij}^{(\tau)} = F^{(t)}\{f[X_i; P; Y_i; Q; a_i^{(\tau)}; a_{ij}^{(\tau)}]\} \tag{2.13}$$

where $X_i = (x_{i1}, x_{i2}, \dots, x_{im})$ and $Q = (q_1, q_2, \dots, q_i)$ are means of specific ponderabilities of the values of the actual levels and the mean of the ponderabilities of manufacturing and field are the most important factors. The study of reliability prediction and the dependence on specific manufacturing and field conditions (operating conditions) of different companies will be discussed later in this book.

Let us build a specific influence function which can help to determine:

- the level of the combined impact of all the most important lack of correlated and generalized factors of manufacture and field action;

- the level of impact of individual groups of factors (the group of manufacturing factors and group of field factors) on product reliability;
- the level of impacts of individual factors of the group of the most important factors.

The functions $f(X_i; P; Y_i; Q; a_i^{(\tau)}; a_{ij}^{(\tau)})$ for all quantities of reliability indexes appear to be equal, because the form of influence for different factors of manufacturing and field is identical for all reliability indexes. But the values for any reliability indexes may be different. The difference should be taken into account with unknown parameters $a_i^{(\tau)}$ and $a_{ij}^{(\tau)}$ which have specific values for each quantity of reliability indexes, for each model of product, each set of field conditions, and for each product of the company.

Let us give for this function the following requirements:

- it must always be positive;
- the maximum value must be less than or equal to 1.00 to give the possibility of simplifying the mathematical model by calculating the functional in Equation 2.14.

In addition, the practical requirement is that the reliability of the specimen used for the testing during the design process is usually higher than after manufacturing of this product, and the time for maintenance is usually less.

Therefore, the coefficients of recalculating the mean time to failure and the mean time of maintenance are characterized by the following dependence:

$$K_{ij}^{(1)} = f(X_i; P; Y_i; Q; a_i^{(1)}) \tag{2.14}$$

$$K_{ij}^{(2)} = 1/f(X_i; P; Y_i; Q; a_i^{(2)}) \tag{2.15}$$

If we take into account the lack of correlation of separate factors and groups of factors, we obtain the following equation (after manipulation):

$$f(X_i; P; Y_i; Q; a_i^{(\tau)}; a_{ij}^{(\tau)}) = C_n \sum_{k-1}^{m} \alpha_x (a_i^{(\tau)})^{1-xik} + C_f \sum_{k=1} \beta_k (a_{ij}^{(\tau)})^{1-yjk} \tag{2.16}$$

where $C_n = 1 - b_n$ and $C_f = 1 - b_r$ are normalized coefficients which relate to the mean specific ponderabilities b_n and b_r of different groups of factors. For this, our research gives the following results: $b_n = 0.47$ and $b_r = 0.53$.

The input of normalized coefficients is necessary, because if ponderability of group of factors (or a separate factor) is greater, the decrease in the product reliability which depends on this group (factor) must be more. It means that the quantity of influence function must be less.

By analogy, normalized coefficients α_k and β_k were included:

$$\alpha_k = \frac{1 - \rho_k}{m - 1}$$

$$\beta_k = \frac{1 - q_k}{l - 1} \tag{2.17}$$

One can find the unknown parameters $\alpha_i^{(\tau)}$ and $\alpha_{ij}^{(\tau)}$ for prototypes, because they cannot be determined for future or modernized products. For example, $\alpha_{ij}^{(\tau)}$ can be evaluated if we compare the τth index of reliability of the ith model of the product which is obtained as a result of ART of μ specimens and as a result of studying of v specimens in the field.

Unknown parameter $\alpha_i^{(\tau)}$ is evaluated by sum of parameters $\alpha_{ij}^{(\tau)}$.

$$\alpha_I^{(\tau)} = N^{-1} \sum_{j=1}^{N} \alpha_{ij}^{(\tau)} \tag{2.18}$$

where N is the number of regions where the previous model is used.

Therefore, for prediction of time to failures the following equation can be recommended:

$$T_{oijf} = T_{oi} \left[C_n \sum_{k-1}^{m} \alpha_k (\alpha_i^{(i)})^{1-xik} + C_f \sum_{k=1}^{i} \beta_k (\alpha_{ij}^1)^{1-yjk} \right] \tag{2.19}$$

where T_{oi} is the mean time to failure. It is obtained a result of ART of μ specimens.

2.5.2.4 Practical Example

As a result of short field testing of new self-propelled spraying machines Ro Gator 554 and John Deere 6500, the mean time to failure and mean time for maintenance were obtained (Table 2.1).

Now let us take as the prototypes the Finn T-90 and T-120. The results of field testing of four specimens of these machines can be seen in Table 2.2.

The values of normalized coefficients α_k, β_k, and q_k were obtained in correspondence with the mean specific ponderabilities of the most important manufacturing and field factors, using Equation 2.16 (Table 2.3).

The unknown parameters of α were obtained using Equation 2.8 and Tables 2.2, 2.3, and 2.4.

The coefficients of recalculating for new machines were obtained (Table 2.5) using the equations and Tables 2.3 and 2.4.

Table 2.1 The results of short field testing of prototypes of self-propelled spraying machines.

	Prototype of Ro Gator 554		Prototype of John Deere 6500	
Index	**Prestige Farms, Clinton (NC)**	**Continental Grain Co. (NY)**	**Prestige Farms, Clinton (NC)**	**Continental Grain Co. (NY)**
Mean time to failure (h)	104	73.80	104	171.10

Table 2.2 Testing results of prototypes of the studied machines.

Conditions of machines used, No. of model	Time to failure (h)
Murphy Family Farm	
Finn T-90	
No. 287	21.9
No. 261	29.04
No. 290	53.62
No. 291	47.81
Mean	37.92
T-120	
No. 059	49.32
No. 030	48.22
No. 063	56.2
No.218	67.41
Mean	58.37
Caroll & Foods	
Finn T-90	
No. 316	40.92
No. 358	1.72
No. 1001	37.67
No. 1005	5871
Mean	39.63
T-120	
No. 714	58.72
No. 1105	80.54
No. 4516	62.98
Mean	67.41

Table 2.3 Normalized coefficients corresponding with the most important manufacturing and field factors.

Normalized coefficient	1	1	3	4	5	6
P_k	0.2325	0.2225	0.2125	0.175	0.1575	—
α_k	0.1920	0.1940	0.1970	0.206	0.2110	—
q_k	0.2325	0.2250	0.2075	0.130	0.1125	0.0925
β_k	0.1540	0.1550	0.1590	0.174	0.1780	0.1820

Table 2.4 Unknown parameters α_i and α_{ij}.

	T-90			T-120		
	Murphy Family Farms	Caroll & Foods	Prestige Farms	Murphy Family Farms	Caroll & Foods	Prestige Farms
$\alpha_i^{(\tau)}$	0.42	0.47	0.445	0.19	0.23	0.21
$\alpha_{ij}^{(\tau)}$	0.45	0.75	0.60	0.56	0.80	0.68

Table 2.5 Coefficient of recalculating for the machines studied.

	Ro Gator 554		John Deere 6500	
	Prestige Farms	Continental Grain Co.	Prestige Farms	Continental Grain Co.
Mean time to failure (h)	0.64	0.66	0.43	0.45

Table 2.6 Predicted mean time to failure.

	Ro Gator 554		John Deere 6500	
	Prestige Farms	Continental Grain Co.	Prestige Farms	Continental Grain Co.
Mein time to failure (h)	57.12	59.01	59.67	62.61

The mean time to failure of the new machines Ro Gator 554 and John Deere 6500 were predicted for when they will be manufactured by series, using Equations 2.22) and (2.23) and Tables 2.1 and 2.5 (Table 2.6).

If one wants to predict system reliability from accelerated testing results of the components, one can use the solution, which was published elsewhere [12].

References

1 Klyatis L. (2012). *Accelerated Reliability and Durability Testing Technology*. John Wiley & Sons.

2 Klyatis L, Klyatis E. (2006). *Accelerated Quality and Reliability Solutions*. Elsevier.

3 Klyatis L. (2016). *Successful Prediction of Product Performance. Quality, Reliability, Durability, Safety, Maintainability, Life Cycle Cost, Profit, and Other Components*. SAE International, Warrendale, PA.

4 Klyatis LM. (2017). Why separate simulation of input influences for accelerated reliability and durability testing is not effective? In *SAE 2017 World Congress*, Detroit, paper 2017-01-0276.

5 Klyatis L, Klyatis E. (2002). *Successful Accelerated Testing*. Mir Collection, New York.

6 Klyatis L, Walls L. (2004). A methodology for selecting representative input regions for accelerated testing. *Quality Engineering* **16**(3): 369–375.

7 Van der Waerden BL. (1956). *Mathematical Statistics with Engineering Annexes*. Springer (in German).

8 Kolmogorov AN. (1941). Interpolation and extrapolation of stationary random sequences. *Izvestiya Akademii Nauk SSSR: Seriya Matematicheskaya* **5**: 3–14.

9 Smirnov NV. (1944) Approximation of distribution laws of random variables by empirical data. *Uspekhi Matematicheskikh Nauk* **10**: 179–206.

10 Ventcel ES. (1966). *Theory of Probability*. Vysshaya Shkola, Moscow.

11 Klyatis LM. (1985). *Accelerated Evaluation of Farm Machinery*. Agropromisdat, Moscow.

12 Klyatis LM, Teskin OI, Fulton JW. (2000) Multi-variate Weibull model for predicting system reliability, from testing results of the components. In *The International Symposium of Product Quality and Integrity (RAMS) Proceedings*, Los Angeles, CA, January 24–27, pp. 144–149.

Exercises

2.1 Describe what happened during Dr. Klyatis's consulting work in improving testing for Black & Decker Company.

2.2 How did Dr. Klyatis continue developing the system he had created earlier for successful reliability prediction for industry after he came to the USA?

2.3 Provide some of the reasons why many publications in reliability prediction could not be successfully used by industry?

2.4 How many basic steps are needed to implement successful reliability prediction? Describe them.

2.5 Why are proper definitions so important in reliability prediction?

2.6 Describe the definition of "accurate system reliability prediction."

2.7 Describe the definition of "correct accelerated reliability testing."

2.8 What basic components need to be simulated for accelerated corrosion testing?

2.9 What basic components are needed to simulate accelerated vibration testing?

2.10 What is the difference between what is commonly referred to as vibration testing and accelerated reliability (or durability) testing?

2.11 What components should be included in the methodology of reliability prediction?

2.12 What components consists of common scheme of methodology for products successful reliability prediction?

2.13 Why is it that many published methodologies cannot be used successfully by industry?

2.14 Consider the basic meaning of criteria of successful reliability prediction using results of accelerated reliability testing.

2.15 Explain the difference between Kolmogorov and Smirnov criteria?

2.16 What is difference between Klyatis criteria and Kolmogorov and Smirnov criteria?

2.17 What is the rule of Klyatis criteria usage?

2.18 What are the basic concepts of reliability prediction using accelerated reliability or durability testing as a source of initial information for prediction calculation?

2.19 What is the basic essence of using the prediction reliability function without finding accurate analytical or graphical forms of the failure distribution law?

2.20 What is the basic essence of predicting with use of the mathematical models without indication of the dependence between product reliability and the different factors of manufacturing and field usage.

2.21 Show the common scheme of successful reliability prediction for industry.

2.22 Demonstrate five common steps for successful reliability prediction. Describe them.

2.23 Show how the interacted groups of real-world conditions for the product/process need to be taken into account for successful reliability prediction.

3

Testing as a Source of Initial Information for Successful Practical Reliability Prediction

Lev M. Klyatis

As was considered in Chapter 1, the application of successful reliability prediction has not yet fully developed for practical use by industry. This is largely because the statistical methodologies that are employed do not directly connect with the sources for obtaining accurate initial information which is a necessary element for the successful application in design of these methodologies. This initial information is usually obtained from testing technologies that are not fully developed to duplicate real-world conditions. Therefore, there is a significant difference between the results predicted and the real-world results during the desired time period, whether this is the service life, warranty period, or any other established time.

3.1 How the Testing Strategy Impacts the Level of Reliability Prediction

Let us consider in greater detail how the testing strategy impacts the level of reliability prediction. There are two basic aspects of testing (Figure 3.1):

- testing in normal field conditions, where the product works in real world;
- accelerated testing.

For reliability evaluation and prediction, one usually uses accelerated testing, for which there are four basic approaches, as shown in Figure 3.2. In this book we used the fourth approach, because:

- The first approach does not provide accurate initial information for successful prediction, because it does not take into account the effect of multi-environmental types of testing influences on the product reliability during the defined life cycle (years of usage).

Reliability Prediction and Testing Textbook, First Edition. Lev M. Klyatis and Edward L. Anderson.
© 2018 John Wiley & Sons, Inc. Published 2018 by John Wiley & Sons, Inc.

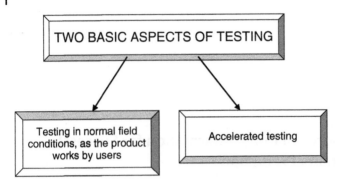

Figure 3.1 Scheme of the two basic aspects of testing.

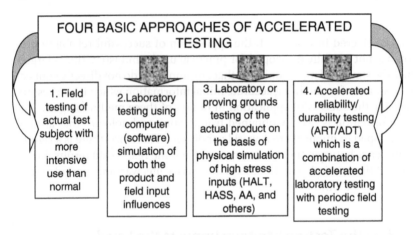

Figure 3.2 Scheme of the four basic approaches to accelerated testing.

- Approach number 2 does not provide reliable information because it does not work with the actual product or the actual influences experienced in the field.
- Approach number 3 does not accurately simulate real field conditions, therefore it does not offer reliable information necessary for successful reliability prediction.

It must be recognized that all types of laboratory testing or proving-ground testing is actually accelerated testing. This is because the objective of the testing is to provide predictive results quicker than can be obtained in normal field conditions. Moreover, accelerated testing usually includes some degree of greater than normal stresses to produce quicker test results.

Figure 3.3 shows the evolution of reliability testing from the traditional separate simulation of the field input influences, which relied primarily on traditional ALT to ART/ADT. It shows how complex the path to accurate ART/ADT truly is.

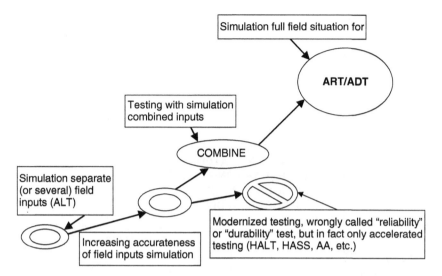

Figure 3.3 The path from traditional ALT with separate (or some) simulation input influences to ART/ADT with simulation for the full field situation (full field input influences plus safety plus human factors).

Figure 3.4 depicts the basic causes of why the first three currently used approaches of accelerated testing usually lead to an unsuccessful prediction.

Figure 3.5 shows that the vast majority of testing that is done is functional testing, followed by a much lower volume of traditional ALT and that combined testing (which began to develop and used in 1950s). ART/ADT is the smallest volume of testing currently in use.

Also highly accelerated life testing (HALT) and the highly accelerated stress screen (HASS) are often improperly referred to as ART or reliability testing. A careful reading of the book by Gregg Hobbs [1], who is the author of HALT and HASS, shows that these types of testing are intensive methods that use stresses higher than those encountered in actual field environments.

It is worthy of note that in that book, Gregg Hobbs never mentions reliability testing, although some people wrongly interpret his approach as being reliability testing (ART or ADT). From Figure 3.6, one can understand the basic causes for the slow development in the area of testing. This figure demonstrates that the majority of technical progress over the last 50–60 years is related to the development of ideas and research.

A lower level of technical progress is depicted for design development, followed by manufacture and service. The lowest level of technical progress relates to testing areas (4). Unfortunately, this is contradictory to what should be the real practice. As new products are introduced with new and higher levels of technology, complexity, and technical progress, their increased complexity requires corresponding higher level of testing if reliability too be successful predicted.

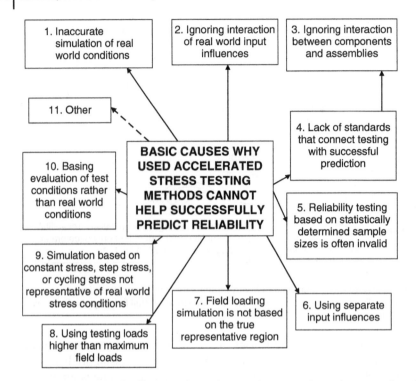

Figure 3.4 Reasons why the currently used approaches to accelerated stress testing often lead to unsuccessful prediction of reliability and durability.

But Figure 3.6 clearly demonstrates the rate of technical progress in testing (curve 4) is increasing much more slowly than the rate of technical progress in ideas, research, and design development. This is one reason for the increase in recalls, product failures, and product liability litigation.

Too often, upper management in organizations is reluctant to invest the funds and other resources needed for testing development. Testing is viewed as a cost center that is to be contained rather than an investment that will return profit through loss avoidance and reduced risk and liability. It is far easier for managers to envision new and innovative product design as the path to investment returns. Unfortunately, as a result, many companies have lost a lot of money in the long term from recalls, complaints, and higher costs, such as redesign or production changes, than was predicted by the design and manufacturing teams. Lev Klyatis previous publications provide examples of this.

Figure 3.6 also helps one understand why three of the four basic approaches of accelerated stress testing currently used frequently lead to inaccurate prediction (1, 2, and 3 in Figure 3.2).

When the initial information used for reliability calculations is poor, the results are likely to be poor. For example, many companies still use single-axis

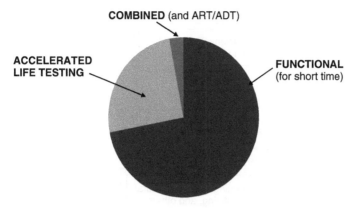

Figure 3.5 Ratio of different areas of testing.

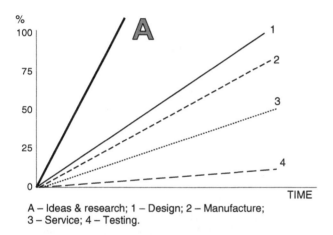

A – Ideas & research; 1 – Design; 2 – Manufacture;
3 – Service; 4 – Testing.

Figure 3.6 Progress of the different areas of activity (over the last 50–60 years).

vibration testing today, as was done for more than the last fifty years. An associated factor is progress in the development of vibration testing equipment also depends on the speed of technical development in the area of electronics which is heavily used in test equipment's systems of control. A second big difference is that what was called vibration testing fifty years ago is now often mistakenly called reliability testing. This can be read about in *Accelerated Reliability and Durability Testing Technology* [2]. A similar situation exists in accelerated corrosion testing. Until recently, only chemical pollution from a single chemical was evaluated, while in the real world there are many chemicals and field conditions present simultaneously, as is shown in Figure 3.9. These are but two of many examples. Figure 3.7 shows the interactions of actual field input influences on failures. Further details can be found elsewhere [2, 3].

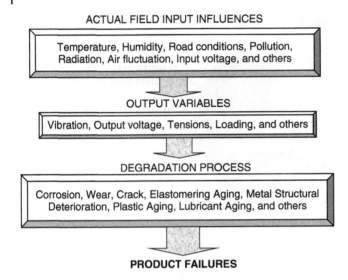

ACTUAL FIELD INPUT INFLUENCES

Temperature, Humidity, Road conditions, Pollution, Radiation, Air fluctuation, Input voltage, and others

OUTPUT VARIABLES

Vibration, Output voltage, Tensions, Loading, and others

DEGRADATION PROCESS

Corrosion, Wear, Crack, Elastomering Aging, Metal Structural Deterioration, Plastic Aging, Lubricant Aging, and others

PRODUCT FAILURES

Figure 3.7 The path from field input influences to failures.

3.2   The Role of Field Influences on Accurate Simulation

For a better understanding of the path depicted in Figure 3.7, it is important to remember the following.

- All types of laboratory and proving-ground testing supposedly duplicate the real-world conditions, but:
  - this testing often does not accurately reflect real-world conditions; and
  - accelerated testing, because it simulates testing faster and/or higher stress levels than in the real world, is also different from actual real-world conditions.
- There are different levels of simulation. Higher level (more accurate) simulation leads to better testing results, and will produce a more successful prediction.
- Step-stress testing is not an accurate simulation, because it generally does not simulate the real conditions of the field situation. Therefore, prediction based on step-stress-testing-indicated reliability may be very different than real-world reliability.

As a result, field use will result in:

- a greater number of failures than was predicted after testing during design and manufacturing;
- more recalls, complaints, and other problems.

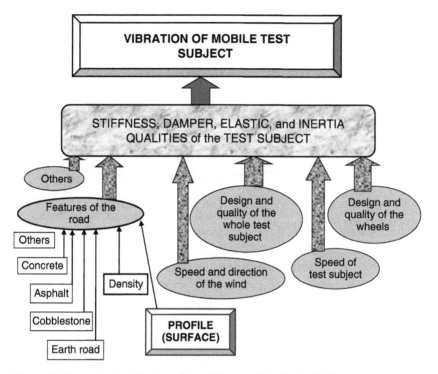

Figure 3.8 Factors in the vibration of a mobile test subject in the field.

As another example, consider commonly used simulations of vibration and corrosion testing. As can be seen in Figure 3.8, while vibration in the field is accompanied by many input influences, usually vibration testing in the laboratory considers only one of these influences—usually the surface profile of the operating terrain. But failure may be a result of other inputs that are present in the actual conditions, and the effects of these other factors then leads to laboratory results different than field results. The final result is unsuccessful prediction. These are the reasons one needs to simulate all relevant influences, as shown in Figure 3.8, for accurate simulation of the field vibration effect of a mobile test subject.

In a similar manner, let us consider accelerated corrosion testing. Corrosion in the field depends on two groups of interconnected influences (Figure 3.9) multi-environmental and mechanical. The multi-environmental group consists of chemical pollution, mechanical pollution, moisture, temperature, and so on. Mechanical influences consist of deformation, vibration wear, friction, and so on. Therefore, for an accurate simulation using an accelerated corrosion process in the laboratory, one needs to simulate the complex input influences. Instead, if one simulates only chemical pollution influences, the predicted results are not likely to be accurate.

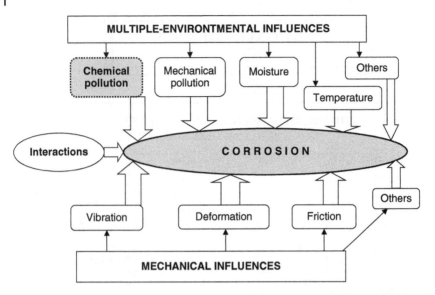

Figure 3.9 Principal scheme of corrosion in the field as a result of multi-environmental and mechanical influences, and their interactions.

The full array of the interacted field input influences on a mobile product is depicted in Figure 3.10, while Figure 3.11 depicts the influences that need to be considered for an accurate simulation of temperature effects.

Further, most products usually consist of a series of interconnected components that, in the real world, interact with each other. Therefore, these component interactions must also be taken into account for accurate simulation, including all the aspects and the full operational range of each input influence, as shown in the example of temperature simulation depicted in Figure 3.11. These can be seen in greater detail elsewhere [2].

For accurate real-world simulation it is necessary to take into account the interactions of not just the unit or product, but also how the unit or product interacts with other elements of the completed functional unit or product, as is shown in Figure 3.12. For example, if the company is designing and manufacturing transmissions for the automotive industry, one needs to take into account that the transmission, first, does not work separately and, second, interacts with the vibrations of the product, as well as other components of the completed car or truck. Unfortunately, companies often do not take this into account.

This is one of the basic causes as to why reliability during stress testing (or other types of ALT) is different from reliability in the field. There are many other problems in using ALT that are not discussed here, (but considered in [2]) and why it is our belief that ART/ADT is a better approach to improving reliability prediction.

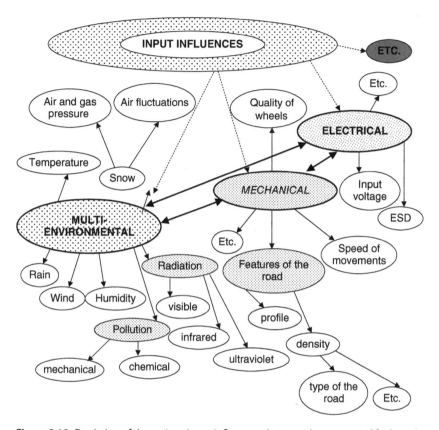

Figure 3.10 Depiction of the various input influences that must be accounted for based on actual field conditions experienced by the product.

3.3 Basic Concepts of Accelerated Reliability and Durability Testing Technology

ART and ADT was described in detail in *Accelerated Reliability and Durability Testing Technology* [2]; therefore, only the basic concepts of this type of testing are detailed here.

ART/ADT consists of two basic components (Figure 3.13). The first basic component, accelerated laboratory testing, as shown in Figure 3.14, includes the interconnected groups of multi-environmental testing, mechanical testing, electrical (electronic) testing, testing for assuring product safety (e.g., crash testing), and so on. These must act simultaneously, as they do in the field. The second component of ART/ADT, periodic field testing, is shown in Figure 3.15.

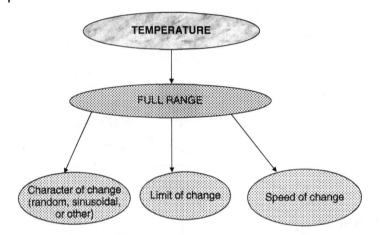

Figure 3.11 Scheme of the study of temperature as an example of input influence on the test subject.

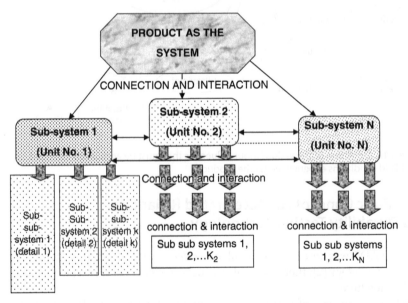

Figure 3.12 The full hierarchy of the complete product and its components as a test subject.

Thus, ART and/or ADT includes all of the following:

- the basic concepts of a strategy for development of accurate physical simulation of the interacted real-world conditions;
- the principles of a system of controls that will provide physical simulation of the random input influences on the actual product;

Figure 3.13 The two basic components of ART/ADT.

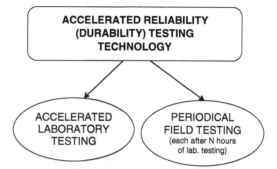

Figure 3.14 Basic components of accelerated laboratory testing as a component of ART/ADT.

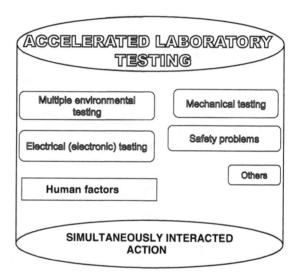

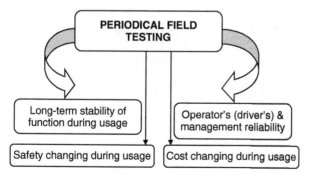

Figure 3.15 Periodic field testing as the second major component of ART/ADT.

- development of the equipment needed for the specific ART of the product;
- control mechanisms for the simulation processes, output variables, and degradation processes;
- managing and conducting ART/ADT for the product by trained and qualified personnel.

It is important for everyone to understand that reliability is not a separate performance component, but it interacts with all of the other performance components, such as durability, maintainability, supportability, life-cycle cost, safety, profit, recalls, and so on, as is discussed further in *Successful Accelerated Testing* [4]. Also, for accurate simulation of field conditions, it is necessary to account for the human factors, such as the way the operator and the way management will actually use the product which often is not necessarily as instructed by the operator's manual, as these also have an impact on the product reliability. These factors are depicted in Figure 3.16.

Further aspects of this issue have been published elsewhere [2–5]. Some components of these concepts can be seen herein. Figures 3.17 and 3.18 demonstrate some of the specific aspects needed to produce accurate physical simulation of field conditions. It must also be noted that this simulation, as well as ART/ADT, is continuing to evolve. The basic trends of these simulations

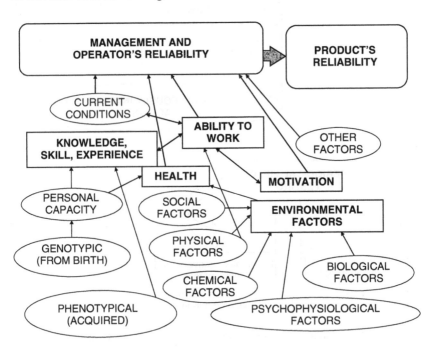

Figure 3.16 Demonstration of the influence of management and operator's reliability on the product/technology reliability [2].

and testing are shown in Figure 3.17, which shows that the basic step is the accurate physical simulation of the field conditions. This is the first basic step one needs for successful reliability prediction. Without a clear understanding of conditions, successful implementation of accurate reliability prediction is not possible.

Once the first step of accurate physical simulation has been completed, the second step is conducting ART/ADT. But, in order to do this, it is also necessary to understand the trends in the development process as shown in Figure 3.17. This development process must be connected closely with the development of an engineering culture (including organizational). Figure 3.18 depicts some of the cultural problems in many organizations that must be overcome to obtain this goal. These are discussed in greater detail in *Successful Prediction of Product Performance* [5].

Unfortunately, some of the reasons for this is that, in the past, professionals in testing and design have not provided accurate simulation of the field performance. As a result, management does not have faith or confidence in providing

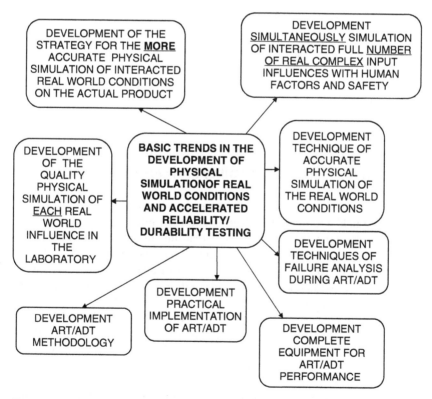

Figure 3.17 Schematic trends in development of physical simulation of the real-world conditions and ART/ADT.

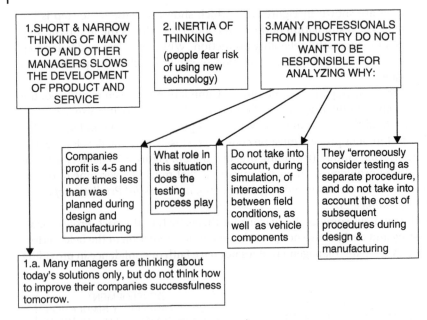

Figure 3.18 Some reasons for low engineering culture.

the necessary backing and resources to improve testing. Better results provided by ART can help to overcome this perception.

3.4 Why Separate Simulation of Input Influences is not Effective in Accelerated Reliability and Durability Testing

It was the many years of practical experience in reliability and durability testing, including field and laboratory testing and their effectiveness, that helped authors to understand the importance of real-life conditions that needed to be simulated in the laboratory for effective reliability predictions. The key element is the need to simulate accurately real-world conditions. This experience clearly demonstrated how real-world conditions interact with each other, and how important it is to take into account these interactions for accurate simulation of the field situation. For example, while it was easily understandable that temperature and humidity do not act separately from pollution, radiation, and other environmental factors, similar interdependencies also exist with electrical, mechanical, and other interacting groups of influences. From this field testing experience and researching the literature about field conditions, it became possible to understand and discern which interactions were meaningful, and which were not.

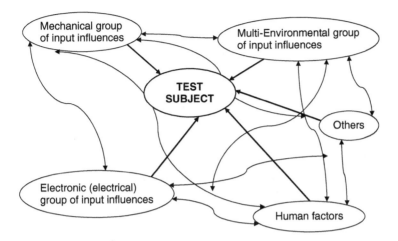

Figure 3.19 Interconnected group of real-world input influences on test subject.

And from experience with the SAE World Congresses, as well as Reliability and Maintainability Symposiums, ASQ World Conferences, IEEE Workshops in Reliability, ASAE Annual International Meetings, and other meetings and publications [6–18], it could be seen that many of the authors of the papers often did not carefully analyze the field conditions. Knowledge of these conditions is needed to accurately simulate in the laboratory, and to obtain accurate information of reliability, quality, and durability of their test subjects in the real world during the product service life or some other specified time or use. (also see the websites of MTS (https://www.mts.com/en/index.htm), K.H. Steuernagel (http://www.aikondocdesign.com/KHS/khspg2.htm), and Arizona Equipment (http://www.awsequipment.com)). The input influences of real-world conditions acting on a test subject are shown in Figure 3.19. As can be seen from this figure, they interact with each other, and the character of these interactions affects product reliability, durability, safety, and other performance measures.

And these interactions relate to all areas: industrial (automotive, aircraft, aerospace, electronic, and so on), medical, social, and so on with the understanding that, for different products, there are different groups and severities of these input influences.

Figure 3.20 depicts the multi-environmental group of input influences. For accelerated testing to be useful as a basis of a test, it needs to accurately simulate the real-world specific input influences shown in Figures 3.20 and 3.21. It is important to remember that in the real world the input influences never act separately, and similar situations exist with product degradation processes in the field.

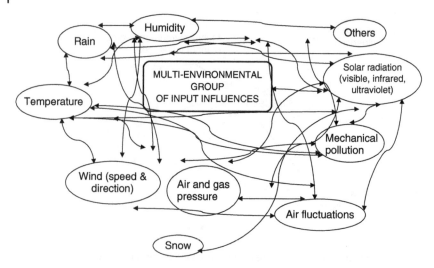

Figure 3.20 Multi-environmental group of input influences.

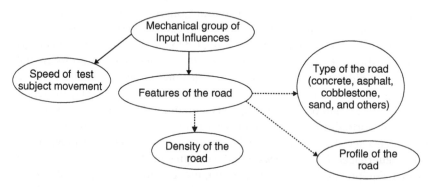

Figure 3.21 Example of content of interacted components of mechanical group of input influences.

The book *Accelerated Reliability and Durability Testing Technology* [2] goes into further detail as to other groups of input influences. Often, there is a misunderstanding of the difference between reliability testing and other types of testing, such as combined, step-stress, vibration, corrosion, and many other types of testing [16–18]. This misunderstanding often leads to significant differences between the laboratory testing results and field performance of the product.

But the basic problem is still that the separate simulation of field input influences is not effective. Let us consider this problem.

It was understandable that, for example, temperature and humidity do not act separately from pollution, radiation, and other environmental factors.

Similar situations exist in electrical, mechanical, and other groups of influences. A second example, that is also understandable, is that the corrosion process is the result of many factors – chemical, mechanical, and so on. This is understandable from both experience and from advanced literature. Field testing experience was an essential element in helping to better understand which literature about field conditions and their actions is reliable and which is not.

An important aspect necessary for the solution of this problem is the correct understanding of the terms and definitions used in this scientific discipline.

For successful simulation and the ability to achieve dependable reliability and durability testing, and ultimately to obtain the true objective successful reliability prediction, it is very important that there be a correct understanding of these terms and definitions. To obtain this goal, the terms and definitions that are included in the referenced books *Accelerated Reliability and Durability Testing Technology* [2] and *Successful Prediction of Product Performance* [5], as well as the standard definitions presented in Chapter 4 of this book, should be used consistently.

From experience, before conducting accelerated testing one should understand the following:

- Worldwide, professionals are conducting accelerated testing in different areas of industry, but the basic problem remains that more accurate simulation of actual field conditions is needed. This problem persists as of the time this book was written.
- For simulation processes to be effective, you need to know precisely what needs to be simulated and how to accurately simulate complicated real field conditions. Clear answers to questions such as what kind of field conditions need to be simulated, and how to evaluate the accurateness of the simulation compared with real world conditions, are needed.
- From experience as a field test engineer, it is important to include the analysis of field conditions acting on the product in written reports of the testing. These reports provided benefits to designers. Often, it is found that such information is not included in reports by colleagues and other test engineers.
- During the preparation of author's PhD dissertation, author's research into improving harvesting technology included experimental research on how changing the influence of different field conditions impacted the product, how they interacted with each other, and, finally, how these interactions influenced the research results. This knowledge came from careful analysis of the real field situation. A major problem is many people involved in research, design, testing, and manufacturing of products do not have enough experience in field testing and research. Too often their only experience is with laboratory research and testing.
- To a similar degree, those involved in the design phase did not know (and often still do not know) enough about the real field conditions, and how

they influence product quality, reliability, durability, and interact with other performance components. Throughout the design, testing, and manufacturing processes, these professionals still spend too little time actually in the field conditions. As a result, they do not understand how important this real experience is to developing a successful product. Figures 3.19 and 3.20 demonstrate examples of the interactions of various field influences.

- Generally, there are several groups of input influences acting simultaneously on the test subject. The input influences of real-world conditions acting on the test subject is shown, as an example, in Figure 3.19. Product reliability, durability, safety, and other performance components are influenced by the level of success of the prediction which depends on the character of such interactions.
- The aforementioned concepts can be applied to all products in all areas: automotive, aircraft, aerospace, electronics, and so on.
- While the overall concepts remain the same, the specifics for different products are often related to different groups of input influences. As an example, Figure 3.20 demonstrates the multi-environmental group of input influences.
- In order for accelerated testing to be useful as a basis of testing, it must accurately simulate the real-world input influences, as shown in Figures 3.20 and 3.21.
- In the real world, input influences never act separately. A similar situation exists with the product degradation process in field operations. For example, the corrosion process in the field is often a result of the interaction of several different types of field input influences.

Figure 3.9 illustrates corrosion as result of interaction of two groups of field input influences: multi-environmental and mechanical. In this illustration, the multi-environmental group consists of the interaction of chemical pollution, mechanical pollution, temperature, humidity, and so on. The mechanical group consists of vibration, deformation, friction, and so on. Accurate simulation of the field corrosion process can only be simulated accurately in the laboratory if these field conditions are accurately simulated.

But many corrosion test chambers available in the market only simulate chemical pollution. This is an example of inaccurate simulation of field conditions for the same basic reason—without accurate knowledge of the field conditions you cannot know what to simulate in the laboratory. This results in inaccurate ART and ADT, which then negatively affects product quality, reliability, durability, and other performance metrics, which then can lead to economic losses, recalls, and other negative impacts of the new products or technologies.

Another element needed to provide an accurate simulation of real-world conditions is understanding that in the real world the test subject (engine, transmission, and so on) is not a discrete and independent unit, but is a component that works in conjunction with other parts of the entire product or the entire machine, as is depicted in Figure 3.12. While all of these input influences interact on the completed machine, too often testing is only done for the part or component.

Unfortunately, many companies who are suppliers that design and manufacture components for a completed machine only test for their part or components. For example, a transmission manufacturer may test their transmission for vibration, but without knowing the effects of the engine and driveshaft as encountered in the real world they do not have an accurate simulation for predicting product reliability. This is especially important for accurate physical simulation of mobile machinery.

Practical implementation of this requires involving a group of engineers and staff who have in-depth knowledge of the interactions of these attached components or a team approach to testing the product with its attachments.

Early in his career, the author was in the position of an engineer and was not able to implement many of these ideas. Later, when the author achieved a management position, he was able to implement these concepts and improve the reliability predictions of the entire unit. It is his firm belief that too many professionals involved in laboratory testing do not understand the need to perform laboratory testing with accurate simulation of the interaction of multiple field influences.

You can see reliability testing being performed in many American, Japanese, European, and other countries based on results obtained from separate vibration testing and separate corrosion testing using test chambers accounting for one influence only. This is also prevalent in climatic (simulation temperature plus humidity) and many other types of testing. There are two basic reasons for this situation:

- First, the perception that the testing and experimental research process will be cheaper and simpler. While this may appear to minimize testing and developmental costs it fails to account for the future losses attributable to the imprecise testing. Unfortunately, these avoidable losses are difficult or impossible to predict and quantify; and, the testing unit is often not deemed responsible for them, as was demonstrated earlier in this book.
- Second, because they do not take into account the influences and interactions that are present in the real world, they never know how well the tested product actually works in the real world. Also, the testing group often does not receive meaningful feedback on a tested product's problems or performance.

As a result, the degradation of the product in the laboratory (during testing) is different from the actual degradation in field use, and too often the final results are complaints, recalls, losses in projected profit, customer dissatisfaction, litigation, and other economic and technology problems.

Unfortunately, this is an all too typical situation resulting from narrow thinking by management. These test systems (ART/ADT technology) provide the ability to find the causative factors that result in low reliability and durability, and thereby avoid or greatly decrease the costs of remediating problems. Consider one example. By accurately simulating in the laboratory the basic field influences, one can study and understand the prevalent degradation processes. And because the testing can use high-speed projecting machines, which expose the product to degradation at a much faster rate than occurs normally, studying and predicting the degradation processes and the ability to make product improvements becomes possible. These processes cannot be done in the field, because the machinery often works in dusty conditions, with a lot of mechanical pollution. It is possible in the laboratory to use many types of apparatus that cannot be used in the field for studying the degradation process. But this can be effective only if one is simulating accurately in the laboratory the field conditions. Many companies do not, or cannot do this.

Chapter 4 details some practical examples of successful implementation reliability testing and prediction. Finally, in this chapter let us briefly consider the cost versus benefit of ART. For this we will use the situation where there is simultaneous combination of different types of input influences, and compare this with single testing, where there is separate simulation of input influences. The cost of the equipment, methodology, and staff time for conducting single tests—for example, separate vibration testing, or corrosion testing, or temperature plus humidity testing, or testing in dust chambers, or input voltage plus vibration testing, and so on—may appear to be less expensive than reliability testing using a combination of all the above types of influences.

This is correct if you only consider the direct costs of the two approaches. But we know the quality of the testing impacts many other costs, such as those related to subsequent processes, including design, manufacturing, usage, service, and so on. For example, the reasons (and costs) for returns, recalls, service bulletins and product modifications are often associated with the poor accuracy of the testing. One must also take into account all the costs associated with future changes in the safety, quality, and reliability during the life of the product. Failure to accurately simulate interacted real-world influences on the product leads to increasing expenses for returns, recalls, and customer confidence and acceptance of the product.

From experience in consulting for many companies, when a test engineer in vibration testing is asked what is the goal and objective of their testing, the usual answer is "reliability." But this answer is wrong; the product's reliability depends on more than the results of vibration testing.

Let us demonstrate an example of poor testing results on the vibration testing of an electronic product. The following example was published in 1999 in the Tutorial Notes of the Annual Reliability and Maintainability Symposium [19]:

In today's stress testing environment, there are two predominate methods for delivering vibration energy to units under test (UUTs): Electrodynamic (ED) shakers and Pneumatic or Repetitive Shock (RS) vibrators, also known as six Degrees of Freedom (DOF) shakers.

ED shakers require a controller to produce an electrical signal, which is fed to a high power amplifier that drives the coil. ED systems are very flexible in that they provide excellent control of the vibration spectrum and can provide sine, swept sine, and shock as well as random vibration. The major disadvantage is their inability to produce multi-axis simultaneous vibration. Some manufacturers overcome this by orienting multiple shakers orthogonal to one another. This configuration is very expensive and generally not applicable to combined environmental testing.

A popular alternative to ED vibration is referred to as 6 DOF or Repetitive Shock vibration. In this application, multiple air driven hammers are mounted to the bottom of a vibration table in orientations that transmit energy in the x, y, and z directions as well as rotational energy about each axis (hence the term six degrees of freedom). Compared to a comparably priced ED system, RS can attain higher g levels and can handle more product mass. The major disadvantage is that the frequency distribution of the vibration spectrum produced by RS cannot be easily controlled. Power sources should be capable of delivering power over the full specified voltage input range of the product under test.

The foregoing relates to the recent time and is only one more example of how low quality of simulation is an obstacle for successful prediction of reliability. Similar situations relate to presentations at the SAE World Congresses, Quality Congresses, and other conferences.

But vibration is only one component of mechanical testing (influences). In the real world the varied mechanical influences act in combination with multi-environmental influences, electrical and electronic influences, and so on. The product's reliability, whether measured by the degradation process, time to failures and between failures, cost of failures, and so on, is a final result of all field input influences in action, and coupled with human factors. Without accounting for these influences one cannot successfully predict the product's reliability.

As a result of poor prediction, during use unpredicted accidents or failures, and increased cost of use can be experienced, which, through the cost of returns, the costs of improvement to the design and manufacturing processes, and other costs such as lost customers, recalls, and litigation produce significant, and unanticipated organizational losses. The book *Successful Prediction of Product Performance* [5] details how, over 30 years, recalls in automotive industry have increased, which leads to increasing losses measured in billions of dollars.

When these costs are considered, it is evident that accurate accelerated reliability (or durability) testing is more economical to the organization than single testing with simulation of only separate influences. Detailed consideration of this problem, with examples, can be found in the book *Accelerated Reliability and Durability Testing Technology* [2]. In conclusion, because of the cost of subsequent problems, single testing with simulation of discrete input influences is more expensive than ART/ADT simultaneous combination, even though the testing cost may be higher.

As demonstrated in Figure 3.3, there is no easy path to transition from traditional ALT, which does not provide the information needed for successful reliability prediction to ART/ADT, which does offer this possibility. This path requires qualitative (accurateness of interactions) and quantitative (increasing number of field inputs) simulation as detailed in the literature [20–24].

References

1 Hobbs GK. (2000). *Accelerated Reliability Engineering: HALT and HASS*. John Wiley & Sons.

2 Klyatis LM. (2002). *Accelerated Reliability and Durability Testing Technology*. John Wiley & Sons, Inc., Hoboken, NJ.

3 Klyatis L, Klyatis E. (2006). *Accelerated Quality and Reliability Solutions*. Elsevier.

4 Klyatis L, Klyatis E. (2002). *Successful Accelerated Testing*. Mir Collection, New York.

5 Klyatis L. (2016). *Successful Prediction of Product Performance. Quality, Reliability, Durability, Safety, Maintainability, Life Cycle Cost, Profit, and Other Components*. SAE International. Warrendale, PA.

6 SAE 2005–2016 World Congress & Exhibition. Event Guides.

7 SAE 2012–2016 AeroTech Congress & Exhibition. Event Guides.

8 ASQ 2006–2016 *World Conferences on Quality and Improvement*. Program.

9 ASQ's 1997–2004 *Annual Quality Congresses*. On-Site Programs & Proceedings.

10 ASAE (The International Society for Engineering in Agricultural, Food, and Technological Systems) 1995–1997. Final Program.

11 SAE 2003–2005 World Aviation Congress & Expositions. Final Program.

12 ASABE (American Society of Agricultural and Biological Engineers). Annual International Meeting. 2007 Program.

13 *Agricultural Equipment Technology Conference* (1999). Program.

14 ASAE (The International Society for Engineering in Agricultural, Food, and Technological Systems) 1994 Winter & *Summer Meetings*. Final Programs.

15 Annual Reliability and Maintainability Symposiums. The International Symposium of Product Quality & Integrity. *Proceedings*, **1997–2002**, 2012.

16 Kyle JT, Harrison HP. (1960). The use of the accelerometer in simulating field conditions for accelerated testing of farm machinery. In *Winter Meeting*, ASAE, Memphis, TN. ASAE Paper No. 60-631.

17 Briskham P, Smith G. (2000). Cycle stress durability testing of lap shear joints exposed to hot–wet conditions. *International Journal of Adhesion and Adhesives* **20**: 33–38.

18 Chan AH (ed.). (2001). *Accelerated Stress Testing Handbook. Guide for Achieving Quality Products*. Wiley–IEEE Press.

19 Chen AH, Parker PT. (1999). Tutorial notes. Product reliability through stress testing. In Annual Reliability and Maintainability Symposium, *January* 18–21, Washington, DC.

20 Klyatis L. (2014). The role of accurate simulation of real world conditions and ART/ADT technology for accurate efficiency predicting of the product/process. In *SAE 2014 World Congress and Exhibition*, Detroit, paper 2014-01-0746.

21 Klyatis LM. (2011). Why current types of accelerated stress testing cannot help to accurately predict reliability and durability? In *SAE 2011 World Congress and Exhibition*, paper 2011-01-0800. Also in book Reliability and Robust Design in Automotive Engineering (in the book SP-2306). Detroit, MI, April 12-14, 2011.

22 Klyatis L, Vaysman A. (2007/2008). Accurate simulation of human factors and reliability, maintainability, and supportability solutions. *The Journal of Reliability, Maintainability, Supportability in Systems Engineering* (Winter).

23 Klyatis L. (2006). A new approach to physical simulation and accelerated reliability testing in avionics. In *Development Forum. Aerospace Testing Expo2006 North America*, Anaheim, CA, November 14–16.

24 Klyatis LM. (1998). Physical simulation of input processes for accelerated reliability testing. In *The Twelfth International Conference of the Israel Society for Quality* (proceedings on the CD-ROM. File://F//Images/Dec982S .htm, pp. 1–12), Jerusalem, December 1–3, abstracts, p. 29.

Exercises

3.1 Show schematically two basic aspects of testing detailed in this chapter.

3.2 Show the scheme of basic approaches to accelerated testing. How many are there?

3.3 Why is only one approach considered in this book?

3.4 Show the scheme of the actual path from traditional testing with simulation of separate or several input influences to ART/ADT.

3.5 Why is HALT or HASS incorrectly called reliability testing?

3.6 Show graphically the rate of technical progress over time comparing the speed of ideas and research, design, manufacturing, service, and testing.

3.7 Why is the speed of technical progress in testing lower than these other areas?

3.8 Show schematically the differences from traditional accelerated testing to ART/ADT.

3.9 List some of the other factors that influence vibration in the real world.

3.10 What factors should be included in laboratory vibration testing? Why is testing only for vibration not accurate?

3.11 List some of the factors that influence corrosion in the field.

3.12 What factor is generally used for simulating corrosion testing in the laboratory. Why is this wrong?

3.13 List some of the input influences on mobile machinery in the field.

3.14 Describe how other components of a complete vehicle connect with each other and must be accounted for in field simulations.

3.15 What are the basic components of ART/ADT?

3.16 What kind of basic components are part of ART/ADT?

3.17 Why is periodic field testing a necessary component of ART/ADT?

3.18 Why are factors such as management and operator reliability needed for product reliability?

3.19 List some of the basic trends in the development of physical simulation of real-world conditions.

3.20 Why is it this author's belief that the lack of an engineering culture is a significant factor in their reluctance to improve testing?

3.21 Why are separate simulation of field influences not effective in ART/ADT?

3.22 List some of the basic groups of input influences acting on mobile products in the field.

3.23 List some of the interacted factors of multi-environmental group of input influences that constitute the field.

3.24 List some of interacted factors that constitute a mechanical group of input influences in the field.

3.25 List some of interacted factors that constitute an electronic group of input influences in the field.

3.26 List some of interacted factors that constitute an electrical group of input influences in the field.

3.27 What costs are frequently overlooked in accounting for the cost of testing?

4

Implementation of Successful Reliability Testing and Prediction

Lev M. Klyatis

While there are different types of implementation, all of them must begin with the students and professionals who will be involved in the implementation learning the new approaches, technologies, strategy, methods, and about the equipment needed for practical implementation. Without first studying, learning, and teaching all those involved in the research, design, manufacturing, service, and usage of the product in the new direction of reliability prediction, ART/ADT, successful implementation will be not possible.

Learning the fundamentals necessary for implementation of new ideas, strategy, approaches, and technologies can be accomplished through:

1. Reviewing publications detailing the new approaches and technologies in the available literature, including the various books, articles, papers, lectures, published presentations, dissertations, reports, protocols, and so on. After this, it is very important to accept the concepts advanced in this literature.
2. Learning about the new approaches and technology from the advanced literature, including this author's lectures, tutorials, and presentations.
3. Using these publications as references in publications of other authors.
4. Citations of the author's basic ideas and research results, with his name, in the books, articles, papers, and project plan, providing a high level of new ideas, approaches, and technologies implementation.
5. Include this knowledge from these approaches and technologies in the preparation of appropriate international, national, and industry standards.
6. Obtain upper level management buy-in by providing presentations, lectures, and tutorials detailing the new approaches, technologies, and the relevant citations with their authors' names, including the benefits to the organization and customers/users.

Reliability Prediction and Testing Textbook, First Edition. Lev M. Klyatis and Edward L. Anderson.
© 2018 John Wiley & Sons, Inc. Published 2018 by John Wiley & Sons, Inc.

7. Develop the new organizational structure, whether through educating existing groups or by forming new teams, divisions, departments, or companies who will be responsible for developing and implementing the new ideas and technologies.
8. Implement the new approaches and technologies in practice, and document the results, demonstrating the improvements gained through the implementation of the new ideas and technology.
9. Document and disseminate the economic gains to the organization achieved by using the new ideas, approaches, and technologies.

Some examples of this will be detailed in this chapter demonstrating successful implementation of new ideas and technology for reliability prediction and testing, described in Chapters 2 and 3 of this book, as well as other appropriate publications and practical experience.

Successful reliability prediction and testing is used in many countries, sciences, and industries over the world. While many of the author's ideas, methods, and equipment have been presented in the author's publications, it is impossible to know all of the successful applications, especially those described in other languages. Therefore, the implementations presented herein will mostly be implementations from the author's practice.

4.1   Direct Implementation: Financial Results

The first implementation of Lev Klyatis's ideas were in Russia (Kalinin State Test Center and company Bezeckselmash) for farm machinery from 1962 through 1965. Later, some of these ideas, methodologies, and new test equipment were implemented in the Ukraine (Figures 4.1 and 4.2), Zelenograd City, Moscow State (Figure 4.3), Israel (Figures 4.4 and 4.5), and Belorussia (Figure 4.6), and then these reliability testing and prediction methodologies were implemented in other countries.

This book's approaches to ART/ADT and reliability prediction have been implemented by many industrial companies. As can be seen by the example in Figures 4.4 and 4.5 (and Tables 4.1 and 4.2), there are economic benefits to these approaches.

Until recently one major obstacle to implementing these new approaches was that many professionals in the areas of design, testing, production, and marketing, both for the companies and their suppliers, worked in isolation without sufficient interaction. They worked with separate requirements (standards) and responsibilities. This is the first problem companies need to overcome for successful implementation of these approaches.

Implementation of these approaches requires a special interdisciplinary team with the following properties:

- The interdisciplinary team needs a chair who is a good manager, who has a good understanding of the different specialties, such as physical and

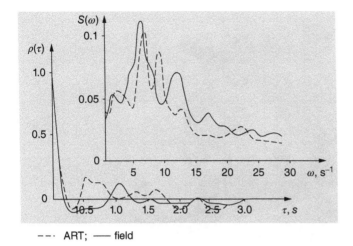

--- ART; —— field

Figure 4.1 Normalized correlation $\rho(\tau)$ and power spectrum $S(\omega)$ of car trailer's frame tension data during ART and field testing (test result from one sensor).

Figure 4.2 Dr. Lev Klyatis, chairman of State Enterprise Testmash and full Professor Moscow University of Agricultural Engineers, in the test center, where implemented his ideas for ART development of farm machinery (1990).

computer simulation, safety, human factors, and so on, and especially knowledgeable of their interconnections and interactions.

- A talented reliability and prediction manager. One who possesses initiative, understanding, and technical capabilities and who will be in charge of the implementation. If necessary, this manager must to be ready to propose and

Figure 4.3 Six-axis vibration test equipment (Testmash, Russia), as a component of ART/ADT. Implemented in Zelenograd Electronic Center, Moscow State (Russia) (1991).

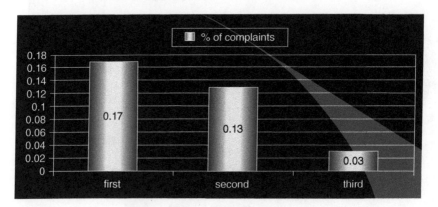

Figure 4.4 Change in the engine complaints for 3 years after implementation of new approach to ART and reliability prediction [1].

organize an interdisciplinary engineering emergency team from the team's professionals. With the support of upper management in research and development, this team will be successful.

- There are many difficulties in the development of this team. Team members must bring good communication skills, an understanding of, and an ability to work cooperatively with each other, as well as the ability to coordinate and appropriately utilize the varied professions and levels of competence of the various team members.

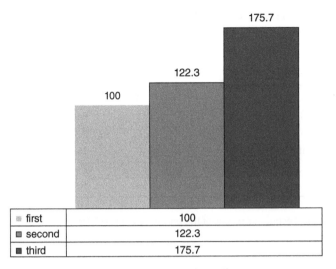

▩ first	100
▣ second	122.3
▪ third	175.7

Figure 4.5 Increased volume of sales of instruments after the implementation of the new approaches, described in Chapters 2 and 3 [1].

Figure 4.6 Dr. Yakhya Abdulgalimov (Testmash) during implementation component of ART/ADT in industrial company Selmash (Bobruysk, Belorussia).

- It is this author's experience that one of the major difficulties of implementing the new approaches is obtaining enthusiasm and cooperation of each team member and maintaining their willingness to work together over time. This is a key requirement for the successful implementation of the new approaches for ART/ADT and successful prediction.
- Another lesson learned in implementing these new approaches is that the team members must include those with an adequate knowledge in the fields of metallic as well as new composite materials, and a high level

Table 4.1 Comparison of the complaints for a reason [1].

	Complaints (%)		
Reason	Year 1	Year 2	Year 3
Design problems	58	43	36
Deviation from the instruction procedure	23	31	34
Not performed according to drawing	10	10	8
Not used according to specification	4	10	15
Others	5	6	7
Total (%)	100	100	100

Table 4.2 Example of practical economic results of the proposed approach for tools [1] by one company.

Year	Sales (10^6 $)	Complaints (no.)	Rejects (%)
2003	134.3	151	3.2
2004	156.2	144	3.0
2005	196.4	127	2.4
2006	214	114	2.0

of understanding in the importance of standardization as a necessary element to reliability and prediction. After implementing these approaches to ART/ADT these companies demonstrated impressive financial improvements.

These methods resulted in increasing the profitability of the companies utilizing them over a 3-year period, with a substantial improvement in the quality and reliability of the product. In fact, one such company that had implemented these approaches and processes was purchased by the billionaire Warren Buffet in Israel in 2006.

This new prediction technology has been implemented by various companies, including those that produce engines, tools, trucks, and other complex products. As a result of implementing these procedures, product sales increased and the reputation of the company in the market place improved (Figures 4.4, 4.5, and 4.7; Tables 4.1 and 4.2).

These figures also demonstrate the importance of establishing the underlying reasons for potential complaints prior to the completion of the design. By correctly implementing accelerated improvement of the product reliability during design, changes can be made early in the design phase and before beginning full-scale manufacturing.

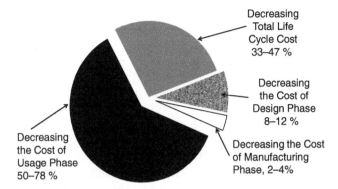

Figure 4.7 Final real results of successful prediction of product reliability.

Ingredients that can be used in improving the final performance of a product can include the quality and the characteristics of the raw materials used, improvement of the components' designs to enhance reliability, improvement of product usefulness to the customers, and others, which may be specific to each manufacturer, depending on the type of final product.

As result of implementing the approaches, explained in Chapters 2 and 3 to ART/ADT technology in various different companies, the following costs and benefits were obtained:

- Decreases to the cost of design phase 10–15%
- Decreases to the cost of manufacture phase 0–2%
- Decreases in the aftermarket cost during use 52–77%
- Total life cycle cost decreases of 33–47%.

Costs in the design and manufacturing can also be minimized by using the results of the ART/ADT in the next models of the product. In many cases at least some of the parts of the production equipment used in the older model can be utilized in the next generation of the product.

If a company does not have the resources for the implementation of ART/ADT technology in one step, it can be done incrementally by first modernizing one test chamber. For example, in vibration testing the next step could be adding equipment for the simulation of temperature and humidity to the same test chamber. The following steps would be adding equipment for pollution simulation; and so forth, until complete modernization has been achieved.

There are many examples of the benefits of ART/ADT and reliability prediction implementation.

4.1.1 Cost-Effective Test Subject Development and Improvement

Often when the ART/ADT results have sufficient correlation with the field results, it is possible to quickly discover the underlying cause(s) for the test subject's degradation and failure. These reasons can be determined by analyzing the test subject's degradation during the time of use and through the determination of the location of the initial degradation and the continued development of the degradation process. This makes it possible to quickly address the reasons for the degradation as determined by ART/ADT. This approach has been used and has proven to be both time saving and cost effective.

If there is not sufficient correlation between test subject degradation in the field and the degradation during ART, the degradation will usually not adequately reflect the experience in the field. If this is the case, the conclusions drawn from ART may not be accurate. In these cases the cost and time needed for refining the test subject and renewed testing to result in better prediction is needed. This must be done rather than attempting to provide accelerated product development without accurate modeling. Unless this is done, the designers and reliability engineers, who often thought they understood the failure (degradation) reasons, changed the design or manufacturing process only to learn that they were wrong in their assumptions and the "improvement" continued to fail in the field. Then they have to revise their work and look for other reasons for the degradation and failure. This situation is found all too often, especially when there is pressure for a rushed solution. Unfortunately, this is a wasteful process that does not allow for proper improvement of the product or the development of proper testing for subject reliability (quality). Moreover, it only increases the cost and time of producing a truly functional product and its protocols for development and improvement. This is a situation familiar in practice.

The following are practical examples that illustrate the aforementioned possibilities of new approach to ART/ADT for the rapid improvement, and prediction of product quality, reliability, and durability.

4.1.1.1 Example 1

An industrial company had problems with the reliability and durability of a model of harvester which they were not able to solve using field testing. They also conducted laboratory testing, but the physical simulation of the field influences in laboratory testing were not accurate. The chief designer of the company asked Lev Klyatis to use his approach for solving the problem. To accomplish this a special methodology of ART/ADT and test equipment were developed to provide accurate prediction of the harvester's reliability during its service life.

For 6 months:

- Two of the original design harvester's specimens were subjected to reliability evaluation for the equivalent of 11 years.
- Three variations of one unit and two variations of another unit were tested. The resulting information was satisfactory on a basis of their service life (8 years) evaluation equivalent.

As results of the finding from this process the reasons for degradation process were eliminated.

- The harvester's design was changed according to the conclusions and recommendations drawn from the testing results.
- Specimens incorporating the design changes were then field tested.
- This reduced the cost (3.2 times) and the time (2.4 times) of the harvester's reliability development.

Reliability validated in actual field use was increased by 2.1 times. Also the design changes for basic components of the harvesters that had previously limited their reliability were confirmed. Usually this work would require a minimum of 2 years for accurate comparisons using present testing methodologies. Accomplishing the above improvement in six months was four times faster.

4.1.1.2 Example 2

There was a problem related to reliability of the working heads, and the special belts that drove them, in another model of the machine. The low reliability of these belts limited the reliability of the whole machine and were an issue beginning with the design stage and continuing through the manufacturing stage, and the field use of the product.

As an attempted remedy, designers increased the strength of the belt, but this only increased reliability by a maximum of 7%, and was accompanied by a doubling in the cost of the belts. The company, Bezeckselmash, asked this author for help in solving this problem. In order to do this, this author's team created test equipment to provide accurate simulation of the field input influences on the belts, corresponding to the methodology described in Chapter 3. By conducting ART of the test subject, corresponding to the methodology described earlier in this book, after several months of testing and studying the reason for the poor reliability of the belts, a reliability prediction was conducted. With the designers, author's team developed recommendations for improving the design of the machinery. The reasons for the belt failures were never in the belt, but in a connected unit. Field testing demonstrated that by implementing these changes the durability of the updated machine was more than doubled, increasing by 2.2 times.

And the cost of this work increased the machine's cost by only 1%, including the cost of the testing equipment and all the work involved in finding the reasons contributing to the failures of the belts.

Figure 4.8 is a drawing of the plan for a test chamber used for reliability testing of trucks by Kamaz, Inc. (Russia, in 1991) which implemented this author's technology. The test chamber design shown incorporates the results of directly implementing author's approach to reliability prediction.

Hence, with accurate initial information from the results of ART/ADT, the methods of reliability prediction can be useful.

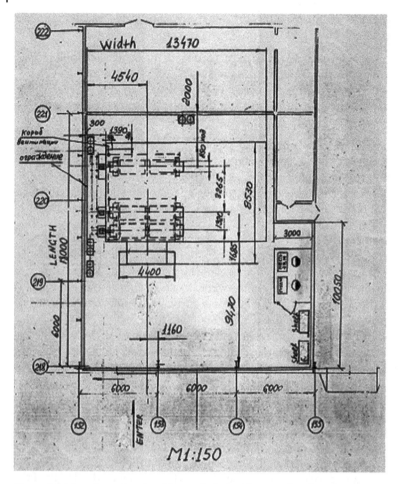

Figure 4.8 Plan of test chamber (Testmash design) for completed truck. Kamaz Inc. (Russia). Engineering Center, Block No. 3.

4.2 Standardization as a Factor in the Implementation of Reliability Testing and Prediction

4.2.1 Implementation of Reliability Testing and Successful Reliability Prediction through the Application of Standard EP-456 "Test and Reliability Guidelines" for Farm Machinery

The implementation of successful reliability prediction began by employing the concept of standardization with farm machinery. We will review how this came about. The author served as a member of American Society of

Agricultural Engineers (ASAE) T-14 Reliability Committee. At the ASAE International meeting in 1995, it was proposed at the ASAE T-14 Committee meeting that the author lead an effort to update the standard EP-456 "Test and Reliability Guidelines" (see Figure 4.9). This effort resulted in updated guidelines including results of research in reliability testing and prediction (see Figure 4.10). These guidelines were approved by the ASAE T-14 Committee and balloted (Figure 4.11) and approved by the ASAE.

Figures 4.9–4.11 demonstrate some of the documents from the updated ASAE standard EP-456 "Test and Reliability Guidelines," which included implementation of new ideas and technology for reliability testing and successful prediction.

4.2.2   How the Work in SAE G-11 Division, Reliability Committee Assisted in Implementing Accelerated Reliability Testing as a Component of Successful Reliability Prediction

Lev Klyatis was also invited to work on the SAE International Society G-11 Division, Reliability Committee. The G-11 Division prepared, discussed, updated, approved, and voted new and current standards related to aerospace reliability, maintainability/supportability, and probabilistic methods, which became SAE standards.

A key objective of SAE G-11 Division was the collaboration in the development of standards. But these efforts were not always successful. The existing SAE draft of standard JA1009 was planned for aerospace reliability testing and prepared earlier, but was not approved in 1998 by the SAE International Aerospace Council.

The SAE G-11 Reliability Committee meeting proposed that a new standard, JA1009 Reliability Testing, should be prepared. Dr. Klyatis made presentations during the G-11 meeting and proposed preparing a group of standards under the common title "Reliability Testing."

This resulted in a group of standards consisting of the following:

- SAE JA 1009/A. Reliability testing – Glossary.
- SAE JA 1009 – 1. Reliability testing. Strategy.
- SAE JA 1009 – 2. Reliability testing. Procedures.
- SAE JA 1009 – 3. Reliability testing. Equipment.
- SAE JA 1009 – 4. Reliability testing. Statistical criteria for a comparison of a reliability testing results and field results.
- SAE JA 1009 – 5. Collection, calculation, statistical analysis of reliability testing data, development recommendations for improvement of test subject reliability, durability, and maintainability.

This group of standards is largely based on results of the development of reliability testing that was described in Chapter 3. This information was

January 20, 1995

TO: ASAE T- 14 Committee Members

FROM: John H. Posselius, Vice Chair

RE: 1994 Winter Meeting Minutes

Greetings fellow committee members. I have enclosed a copy of the minutes from our winter meeting. If there are any problems recorded in the minutes please advise me at your earliest convenience. I will wait a couple of weeks before I send a copy of the minutes to headquarters for their records.

I have spoken to Russ Hahn (at headquarters) about the rewrite of our EP 456. He will send a copy of EP 456, on disc, to either me or directly to Lev Klyatis, along with procedures for rewriting a standard. I indicated that Lev should be listed as the coordinator of the project and that the rewrite should be listed as active.

I am also in the process of sending a copy of the proposed check list for standards to headquarters with the instructions that our committee is willing to review any new (not rewrite) standards as deemed necessary.

Ford New Holland, Inc.
_____
500 Diller Avenue
P O Box 1895
New Holland PA 17557-0903
Telephone (717) 355-1121

Figure 4.9 The letter from ASAE T-14 Committee that Lev Klyatis should be listed as the coordinator of the project "Rewriting the Standard EP 456 Test and Reliability Guidelines."

UNAPPROVED DRAFT 8/97
FOR REVIEW ONLY

ASAE Engineering Practice: ASAE EP456

THIRD DRAFT
PREPARED BY Dr. LEV M. KLYATIS

TEST AND RELIABILITY GUIDELINES

Developed by the ASAE Testing and Reliability Committee

SECTION I—PURPOSE AND SCOPE

1.1 This Engineering Practice shows how product life can be specified in probabilistic terms, how life data should be analyzed, and presents the statistical realities of life testing, including a random failure rate rate.

1.2 The main purpose of the Engineering Practice is the evaluation of reliability measures and identification of test procedures that provide the information for this evaluation.

1.3 This Engineering Practice is not intended to be a comprehensive study on the large variety of subjects encompassed in the analysis of life testing, but rather a guide to demonstrate the power, usefulness and methodology of reliability analysis.

SECTION 2—BACKGROUND

2.1 A key characteristic of a product is its life expectancy. In the agricultural equipment industry a great deal of testing is conducted on materials, components and complete machines to determine expected life.

2.2 Product life needs to be specified in probabilistic terms. This mandates, that test programs require enough replications to minimize the risk of erroneous test conclusions. The test engineer and designer must establish a test program to balance test time and cost against risk of a product failure, further product improvements, and schedule costs.

2.3 The primary emphasis is on the problem of quantifying reliability in product design and testing.

2.4 The only way to measure system reliability is to test a completed product with a combination of influences of components, under conditions that simulate real life, until failure and degradation occurs. One simply cannot assess reliability without data, and the more data available, the more confidence one will have in the estimated reliability level. Extensive testing is often considered undesirable, because it results in expenditures of time and money. Thus, one must consider the trade-off between the value of more confidence in estimated reliability versus the cost of more testing. It is easy to arrive at a test program's cost; the cost of not having the test program is difficult to calculate.

2.5 Before the reliability of a test subject is measured, certain procedural constraints must be established. For example, the amount of preventive maintenance permitted, if it is permitted at all, and the degree to which the system operator can participate in correcting failures must be specified, because the manner in which the system is operated can affect the calculated reliability level.

2.6 If the particular failure density and the distribution function are known, the reliability function can be found directly. The distribution of failure should consider different types of distribution (normal , Weibull , exponential , log normal , gamma , Student , Rayleigh , etc.) In practice, one is usually forced to select a distribution model without having enough data to actually verify its appropriateness. Therefore the procedures used to estimate the various reliability measures may be obtained from empirical data.

2.7 Types of Test for Reliability Testing and Planning: Functional Tests, Environmental Tests, Reliability Life Tests.

2.8 This Engineering Practice is used in order to raise the level of knowledge and skill of test engineers and researchers who work in the area of testing and reliability, and increase the quality and productivity of machinery, thereby decreasing cost of design and test work. This is especially crucial in the following areas: reliability measures and how to evaluate; environmental tests and the influence on the reliability; accelerated testing and evaluating of agricultural equipment; physical simulation of the life rates in the laboratory, and grounding of test regimes; testing not only details and units, but complete machines and equipment.

2.9 The strategy of test for the evaluation of farm machinery reliability include: a) reception of initial information; b) simulation of operating conditions; c) accelerated testing; d) statistical applicants; e) reliability evaluation.

2.10 Testing and evaluation of reliability can be for different levels of machinery life: 1) for one work season; 2) for optimal product life; 3) for service life; 4) on the instruction of the buyers. The machinery can be tested in practice under ordinary operating conditions, laboratory conditions, and combinations of both these types of testing (Fig. 1.). Tests under laboratory conditions consist of (Fig. 1.) tests on the mechanical and environmental influences.

SECTION 3—TERMINOLOGY

3.1 Distribution: A mathematical function giving the cumulative probability that a random quantity, such as a component's life, will be less than or equal to any given value.

3.2 Probability: Likelihood of occurrence based on significant tests.

3.3 Sample: A small number of which will be considered as representative of the total population.

3.4 Accelerated test: A test in which the deterioration of the test subject is intentionally accelerated over that expected in service.

3.5 Accuracy: A generic concept of exactness related to the closeness of agreement between the average of one or more test results and an accepted reference value or the extent to which the readings of a measurement approach the true values of a single measured quantity.

3.6 Characteristic: A property of items in a sample population which, when measured, counted or otherwise observed, helps to distinguish between the items.

3.7 Control (evaluation): An evaluation to check, test, or verify.

3.8 Standard deviation: The most usual measure of the dispersions most Often to form an estimate of some population parameter.

3.9 Population: The totality of items or units of equipment under

1

Figure 4.10 The first page of the third draft of the standard EP 456 "Test and Reliability Guidelines."

discussed and later published at the SAE 2013 World Congresses (papers #2013-01-0152 "Development Standardization 'Glossary' and 'Strategy' for Reliability Testing as a Component of Trends in Development of ART/ADT" [2] and #2013-01-151 "Development of Accelerated Reliability/Durability Testing Standardization as a Components of Trends in Development Accelerated Reliability Testing (ART/ADT)" [3].

COOPERATIVE STANDARDS PROGRAM

5 November 1997

To: **T-14, Testing and Reliability Committee**

R. Steven Newbery, Chair	Lev M. Klyatis
Linwood H. Bowen	John H. Posselius Jr
Lawrence H. Ellebracht	David J. Sandfort
David D. Jones	Frank E. Woeste

Inf cc: James A. Koch, Chair PM-03

From: Dolores Landeck, Standards *Dolores Landeck*

Re: Proposed revision of EP456, Test and Reliability Guidelines

You may recall that development of X456 had been impeded because of language difficulties. Project leader Lev Klyatis subsequently obtained assistance with a rewrite, and the resulting draft is enclosed for your review.

I understand that most of you are well acquainted with the development of this revision. Nevertheless, for your reference, I am including various committee records, provided by Dr Klyatis, which offer some details of the project.

For those of you unfamiliar with the balloting process, the initial ballot period runs 30 days; a reminder will then be sent to anyone who hasn't returned a ballot, extending the review period for another 15 days. It is hoped that by the end of that time, at least 80% of the committee will have responded; if not, a final notice will be sent specifying the final closing date for the ballot. The draft will be considered approved if, after deducting Waive votes, three quarters of the committee approves.

Please review the enclosed materials and return your ballot no later than **5 December**. Your prompt response will be appreciated.

Thank you!

ASAE
2950 NILES ROAD • ST. JOSEPH. MI 49085-9659 • USA • 616/429-0300 • FAX: 616/429-3852 • E-MAIL: HQ@ASAE.ORG

Figure 4.11 The ballot for the ASAE standard EP 456 (Lev Klyatis, project leader).

The SAE Reliability Committee approved the first standard, SAE G-11 JA1009/A Reliability Testing – Glossary, and recommended integrating this with existing standard ARP 5638.

The second standard, SAE JA1009 – 1 Reliability Testing – Strategy, was prepared, discussed, and approved for balloting, but the G-11 Reliability Committee became inactive, and final balloting was not completed.

Drafts of both of these standards were included in the book *Successful Prediction of Product Performance* [4].

From the preparation of these standards, and as a result of discussions on G-11 meetings, the author's theories were updated on ART/ADT that had been developed earlier. The testing methodology developed in these new publications was a key factor for successful prediction of product reliability and other components of the product performance. The presentations and publications of the essence of these standards were included in papers for the SAE World Congresses in Detroit, journal articles, and other publications in this area, and they have proven to assist everyone who wishes to implement successful prediction of product reliability. As a result of development of these standards and career work in quality and reliability, L. K. (Lev Klyatis) was invited to work as a consultant and seminar instructor for several companies and organizations. As a part of G-11 meetings, committee experts often visited companies in geographic areas near the meeting. Figure 4.12 shows the members during a visit to the NASA Langley Research Center.

Figure 4.12 Lev Klyatis (second from left) with group of experts from SAE International, G-11 Division in NASA Langley Research Center, NASA.

These meetings helped this author better understand the methods in use at NASA and companies involved in aerospace, and their understanding of how to obtain successful prediction of product performance. It was surprising to learn that it was difficult to find the person or group at the Langley Research Center who was responsible for reliability testing, and, from what could be learned, their reliability testing professionals were not directly involved in reliability and durability testing of the components or the completed product. Some of the documents related to these standardization efforts are presented later in this chapter.

Figure 4.13 shows the group of SAE G-11 Division during the meeting in Washington, DC, in 2012. At this meeting L. K. (Lev Klyatis) gave a presentation for the G-11 Division members that helped them better understand the concepts for successful reliability prediction. For example, the discussion with Mr. Dan Fitzsimmons, The Boeing Company Technical Fellow, Statistical QE Supplier Management & Procurement Commercial Airplanes, helped him to understand how the structure of large companies makes it difficult to implement these methodologies for improving reliability and durability testing. Organizational structures often involve divisional hierarchies where each vice president or manager has control and responsibility only for their direct areas. And, as we already learned, successful prediction of product performance requires these interactions to be successful. A product's success is not related to a single corporate area, but depends on the efforts of all groups from basic areas of research, testing, design, production, management,

Figure 4.13 SAE G-11 Division members in standardization of reliability and maintainability in aerospace, during a meeting in Washington, DC. Lev Klyatis is fourth from the left.

marketing, and so on that influence the product life-cycle cost, profit, recalls, and other performance components.

The lessons learned from these interactions were included in work for G-11, including several presentations on reliability testing for the G-11 Division and preparation of the aforementioned six SAE standards under the common title "Reliability Testing."

Box 4.1 gives an example of the meeting agenda with the presentation "Overview of the Role and Contents of Six G-11 JA1009 Reliability Testing Standards."

Box 4.1 SAE G-11, Reliability, Maintainability/Supportability and Probabilistic Methods Systems Group Meeting Agenda

November 28–29, 2012. Washington, DC

Hosted by:
Honeywell International 101 Constitution Ave, Washington, DC 20001

Day 1: Wednesday–November 28, 2012

10:00–10:30	Coffee and greetings
10:30–10:45	Host welcome – *Honeywell (speaker: TBD)*
10:45–11:00	Agenda review
11:00–11:15	G-11 Division Overview: prior work, charter and discussion of go-forward plans - *Michael Gorelik*
11:15–11:30	SAE staff presentation – *Donna Lutz, SAE Aerospace*
11:30–12:00	Review of Sept. 2011 meeting minutes
12:00–1:00	*Lunch Break*
1:00 - 1:45	CBM Recommended Practices Guide Project status (Including RAMS-5 meeting and DoD Maintenance Symposium panel session debriefs) – *Chris Sautter and JC Leverette*

1:45–2:30	Review of the status and plans by committee:
	  – *G-11 Reliability*
	  – *G-11 Maintainability/Logistics*
	  – *G-11 Probabilistic Methods*

Day 2: Thursday –November 29, 2012

8:30–8:45	Review of Day 2 Agenda
8:45–9:30	**"Overview of the Role and Contents of Six G-11 JA1009 Reliability Testing Standards"** *– Lev Klyatis*

9:30–12:00	Committees breakout sessions (continued)
	  – *G-11 Reliability*
	  – *G-11 Maintainability/Logistics*
	  – *G-11 Probabilistic Methods*

(Continued)

Box 4.1 (Continued)

12:00–1:00   Lunch Break

1:00–3:30   Committees breakout sessions (continued)

      – *G-11 Reliability*

      – *G-11 Maintainability/Logistics*

      – *G-11 Probabilistic Methods.*

Following the presentation the second standard was prepared, discussed, and approved for voting. But it was not adopted because final voting had not yet occurred, and thereafter the G-11 Reliability Committee stopped all work for several years.

Fortunately, this work resumed in 2017. Box 4.2 contains the SAE G-11 Spring 2017 meeting announcement.

Box 4.2 SAE G-11 Reliability Maintenance Support and Probabilistic Methods Committees Meeting (2017)

StandardsWorks@sae.org   Jan 6 (1 day ago)

to me

Hi All

I am please to announce that we will be having a G-11 Group Meeting on Tuesday 24th Jan in-conjunction with the RAMs conference in Florida.

Please see attached meeting notice for meeting times and location, if you will be attending it will be greatly appreciated if you can register via standard works. I also have attached the list of documents that are up for review.

Please let anyone know that would be interested on working with the G-11 committees and there standards.

If you require further information please contact me on sonal.khunti@sae.org

Kindest Regards
Sonal Khunti

2 Attachments

Preview attachment G-11 Committees Documents to be Reviewed .docx

View attachment G-11 Spring 2017 Meeting Notice.doc

G-11 Spring 2017 Meeting Notice.doc

| StandardsWorks@sae.org | **Hi All I am please to announce that we will be having a G-11 Group...** | Jan 6 (1 day ago) |

Documents to be Reviewed for Each Committee
G-11 Probabilistic Methods Committee

Document List

Document	Title	Date	Status
AIR5080	Integration of Probabilistic Methods into the Design Process	May 08, 2012	Reaffirmed
AIR5086	Perceptions and Limitations Inhibiting the Application of Probabilistic Methods	May 08, 2012	Reaffirmed
AIR5109	Applications of Probabilistic Methods	May 08, 2012	Reaffirmed
AIR5113	Legal Issues Associated with the Use of Probabilistic Design Methods	May 08, 2012	Reaffirmed

G-11R, Reliability Committee

Works in Progress

Project	Title	Sponsor	Date
ARP6204	Condition Based Maintenance (CBM) Recommended Practices	Fred Christian Sautter	Sep 15, 2011
JA1009	*Reliability Testing Standard*	*Lev Klyatis*	*Oct 27, 1998*
JA1009-1	*Reliability Testing - Strategy*	*Lev Klyatis*	*Apr 26, 2013*

(Continued)

Box 4.2 (Continued)

G-11 Reliability Applications Committee

Document List

Document	Title	Date	Status
ARP5638	Rms Terms and Definitions	Mar 06, 2005	Issued

Meeting Location: Rosen Plaza Hotel; 9700 International Drive. Orlando FL, 32819

Meeting Time: 1.30pm – 4.00pm; Meeting Room: Salon 17

If you have any questions, please contact Sonal Khunti, Aerospace Standards Specialist, at +44 7590184521 or sonal.khunti@sae.org.

In conjunction with RAMs Conference. Meeting: January 24th, 2017

Resulting from the discussion during the SAE International G-11 Reliability Committee meetings, the following proposals concerning the JA 1009 Reliability Testing Standard were recommended:

1. *Reliability Testing Standard JA 1009/A: Glossary (of terms and definitions).* See Box 4.3. Consists of a description of the applicable documents (SAE publications, ECSS publications, ASO and federal publications, IEC publications, US Governmental publications, other publications, applicable references, and Glossary of terms and definitions: accelerated reliability testing, accelerated durability testing, durability testing, reliability testing with appropriate notes, accurate simulation of field input influences, accurate prediction, accurate system of reliability prediction, an accurate physical simulation, classification accuracy, correlation, hazard analysis, hazard control, hazard reduction, human factors, human factors engineering, multi-environmental complex of field input influences, output variables, compliance test, laboratory test, accelerated testing, reliability, reliability improvement, reliability growth, durability, failure, failure mechanism, time to failure, time between failures, item, fault, system, subsystem, life cycle, useful life, characteristic, simulation, prediction, life cycle costing, special field testing, and so on (total 11 pages).

2. *Reliability Testing Standard JA 1009/2: Procedures.* Reliability testing (ART/ADT) as a key factor for accurate prediction quality, reliability, durability, maintainability, life-cycle cost, and profit during given time, and accelerated product development. Consists of a description of the step-by-step procedure for the design of reliability testing which is intended for any specific equipment to be tested, when it is considered necessary to simulate closely the real conditions of use of the test subject. It applies fully to laboratory testing in combination with special field testing, and so on.

3. *Reliability Testing Standard JA 1009/1: Reliability Testing – Strategy.* See Box 4.4. Consists of basic concepts of reliability (accelerated reliability/durability) testing, accuracy of the simulation of field conditions (input influences in interconnection with safety, and human factors), physical (chemical) degradation mechanism of the test subject as a criterion for an accurate simulation of field input influences, requirements to obtaining accurate initial information from the field, methodology for selecting a representative input region for accurate simulation of the field conditions, moving the field to the laboratory, and so on (total 18 pages).

4. *Reliability Testing Standard JA 1009/3: Reliability Testing – Equipment.* Consists of requirements for equipment for accelerated reliability testing and its components (multi-environmental, vibration, dynamometer, wind tunnel, etc.). Test equipment considered as a combination of multi-environmental + mechanical + electrical (electronic) and other types of testing. Examples of equipment for reliability/durability testing in the laboratory, and so on.

5. *Reliability Testing Standard JA 1009/4: Statistical criteria for a comparison of reliability testing results and field results.* Consists of statistical criteria for comparison *during* reliability testing of the output variables and physics of degradation process with the output variables and degradation process in the actual service conditions, statistical criteria for comparison *after* reliability testing of the reliability indexes (time to failure, failure intensity, etc.) with these indexes in actual service, and so on.

6. *Reliability Testing Standard JA 1009/5: Collection, calculation, statistical analysis of reliability testing data, development recommendations for improvement of test subject reliability, durability, and maintainability.* Consists of methodology: data collection during the test time, statistical analysis of these data, methods for analysis of the reasons of degradation and failures, preparations recommendations for eliminations of these reasons, counting the accelerated coefficient, development recommendations for improving the reliability of test subject.

Box 4.3 Draft of SAE International Standard Reliability Testing—Glossary

SAEInternational	SURFACE VEHICLE/ AEROSPACE RECOMMENDED PRACTICE	**SAE** JA1009 PropDft XXX2012
		Issued Proposed Draft 2012-04-24
	Reliability Testing—Glossary	

RATIONALE

JA1009, Reliability Testing, is being revised to broaden the technical content and to better organize the material.

FOREWORD

This standard defines words and terms most commonly used which are associated with Reliability Testing (Accelerated Reliability Testing). It is intended to be used as a basis for reliability testing definitions and to reduce the possibility of conflicts, duplications, and incorrect interpretations either expressed or implied elsewhere in documentation. The Reliability Testing Standard corresponds to SAE International Fact Sheet: SAE Technical Committee G-11 Reliability, Maintainability, and Probabilistic Methods, Standards development/revision activities.

Terms and their definitions included in this standard are:

1. Important in acquisition of weapon systems for precise definition of reliability testing (including accelerated reliability and durability testing) criteria.
2. Unique in their definitions, allowing no other meaning.
3. Expressed clearly, preferably without mathematical symbols.

Terms that were not included in this standard are:

1. Found in ordinary technical, statistical, or standard dictionaries or texts having a single acceptable meaning when used in the relevant context.
2. Terms already existing in other standards outside of the project scope.
3. Multiple word terms unless needed for uniqueness.

1. SCOPE

This document applies to reliability testing that is performed to support aerospace applications.

1.1 Purpose
The purpose of this standard is to define words and terms used most frequently in specifying Reliability Testing (Accelerated Reliability and Durability Testing) to give these terms a common meaning for the contractors and users in aerospace.

2. REFERENCES

2.1 Applicable Documents
The following documents form a part of this document to the extent specified herein. The latest issue of SAE publications shall apply. The applicable issue of other publications shall be the issue in effect on the date of the purchase order. In the event of conflict between the text of this document and references cited herein, the text of this document takes precedence. Nothing in this document however, supersedes applicable laws and regulations unless a specific exemption has been obtained

2.1.1   SAE Publications
Available from SAE International, 400 Commonwealth Drive, Warrendale, PA 15096-0001, Tel: 877-606-7323 (inside USA and Canada) or 724-776-4970 (outside USA), www.sae.org.
   ARP 5638 RMS Terms and Definitions.

2.1.2   ECSS Publications
Available from European Cooperation for Space Standardization (ECSS).
   ECSS Secretariat, ESA ESTEC, P.O. Box 299, 2200 AG Noordwijk, The Netherlands, Phone: +31-71-565 5748 Fax: +31-71-565 6839, ecss-secretariat@esa.int

ECSS-Q-30B	European Cooperation for Space Standardization (ECSS).
	Space Product Assurance. Dependability.
ECSS-Q-30B	Glossary of Terms.

2.1.3   ISO and Federal Specifications
ISO 9000:2000 Quality Management Systems—Fundamentals and Vocabulary.

2.1.4   IEC Publications
Available from International Electrotechnical Commission, 3, rue de Varembe, P.O. Box 131, 1211 Geneva 20, Switzerland, Tel: +44-22-919-02-11, www.iec.ch.
   IEC 60050-191:1990 Quality Vocabulary—Part 3: Availability, Reliability and Maintainability terms—Section 3.2, Glossary of International terms.
   IEC 60050-191 International Electrotechnical Vocabulary—Chapter 191: Dependability and Quality in Service (see http://www.electropedia.org/iev/iev.nsf/index?openform&part=191)

2.1.5   US Government Publications
Available from the Document Automation and Production Service (DAPS), Building 4/D, 700 Robbins Avenue, Philadelphia, PA 19111-5094, Tel: 215-697-6257, http://assist.daps.dla.mil/quicksearch/

(Continued)

Box 4.3 (Continued)

MIL-STD-280	Definitions of Item Levels, Item Exchangeability, Models, and Related Terms.
MIL-STD-721C	Definitions of Terms for Reliability and Maintainability.
MIL-HDBK-781	Reliability Test Methods, Plans and Environments for Engineering Development, Qualification, and Production.
MIL-STD-882	System Safety Program Requirements.

2.1.6 Other Publications

Chan, H. Antony, T. Paul Parker, Charles Felkins, Antony Oates, 2000, *Accelerated Stress Testing*. IEEE Press.

Klyatis Lev M., 2012, *Accelerated Reliability and Durability Testing Technology*, John Wiley & Sons, Inc.

Klyatis, Lev M., Eugene L. Klyatis, 2006, *Accelerated Quality and Reliability Solutions*. Elsevier, UK.

Nelson, Wayne, 1990, *Accelerated Testing*. John Wiley & Sons, New York, NY.

Reliability Toolkit, Commercial Practices Edition. Reliability Analysis Center. 1993.

2.2 Applicable References

The Chicago Manual of Style, 14th Edition, University of Chicago Press, Chicago, IL, 1993.
   Webster's Ninth New Collegiate Dictionary.

GLOSSARY OF TERMS AND DEFINITIONS

Accelerated testing	Testing in which the deterioration of the test subject is accelerated.
Accelerated reliability testing or accelerated durability testing (or durability testing)	Testing in which: (a) The physics (or chemistry) of the degradation mechanism (or failure mechanism) is similar to this mechanism in the real world using given criteria and (b) The measurement of reliability and durability indicators (time to failures, its degradation, service life, etc.) has a high correlation with these indicators measurement in real world using given criteria.

NOTE 1	Accelerated reliability testing, or accelerated durability testing, or durability testing offer useful information for accurate prediction reliability and durability, because they are based on accurate simulation of real world conditions.
NOTE 2	It is identical to reliability testing if reliability testing uses for accurate reliability and durability prediction during service life, warranty period, or other.
NOTE 3	Accelerated reliability and durability testing, such as accelerated testing, is connected with the stress process. Higher stress means a higher acceleration coefficient (ratio of time to failures in the field to time to failures during ART), lower correlation between field results and ART results, less accurateness of prediction.
NOTE 4	Accelerated reliability and accelerated durability testing (durability testing) consists of:
	A complex of laboratory testing and periodic field testing. The laboratory testing that provides a simultaneous combination of a whole complex of multi-environmental testing, mechanical testing, electrical testing, and other types of real world testing. The special field testing that takes into account the factors which cannot be accurately simulated in the laboratory, such as the stability of the product's technological process, how the operator's reliability influence on test subject's reliability and durability. Accurate simulation of field conditions requires full simulation of the field input influences integrated with safety and human factors.
NOTE 5	Accelerated reliability testing (ART) and accelerated durability testing (ADT) (or durability testing) have the same basis – the accurate simulation of the field situation. The only difference is in the indices of these types of testing and the length of testing. For reliability it is usually the mean time to failures, time between failures, and other parameters of interest; for durability it is amount of time or volume out of service.
NOTE 6	Accelerated reliability testing can be for different lengths of time, i.e. warranty period, one year, two years, service life, and others.
Acceptance of risk	Decision to cope with consequences, should a risk scenario materialize.
Accurate prediction	Is possible if one has:
	(a) Prediction methodology to incorporate all active field influences and interactions;
	(b) Accurate initial information from accelerated reliability/durability testing.

(Continued)

Box 4.3 (Continued)

Accurate simulation of field influences	If all field influences act simultaneously and in mutual combination, *the field input influences* and simulated accurately.
Accurate system of reliability prediction	The system of prediction is accurate if, and only if, the simulation is accurate and accelerated reliability testing is possible.
Allowable stress	Maximum stress that can be permitted in a structural part for a given operating environment to prevent rupture, collapse, detrimental deformation or unacceptable crack growth.
An accurate physical simulation	Occurs when the physical state of output variables in the laboratory differs from those in the field by no more than the allowable limit of divergence.
Assessment	Any systematic method of obtaining evidence from tests, examinations, questionnaires, surveys and collateral sources used to draw inferences about characteristics of people, objects, or programs for a specific purpose.
Certification	Procedure by which a third-party gives written assurance that a product, process or service conforms to specified requirements.
Classification accuracy	The degree to which neither false positive nor false negative categorizations and diagnoses occur when a test is used to classify an individual or event.
Common cause failure	Failures of multiple items occurring from a single cause which is common to all of them.
Common mode failure	Failures of multiple similar items that fail in the same mode.
Common mode fault	Faults of multiple items which exhibit the same fault mode.
Confidence interval	An interval between two values on a score scale within which, with specified probability, a score or parameter of interest lies.
Configuration control	Activities comprising the control of changes to a configuration item after formal establishment of its configuration documents.
NOTE	Control includes evaluation, coordination, approval or disapproval, and implementation of the changes.
Configuration identification	Activities comprising determination of the product structure, selection of configuration items, documenting the configuration item's physical and functional characteristics including interfaces and subsequent changes, and allocating identification characters or numbers to the configuration item and their documents.

Configuration item	Aggregation of hardware, software, processed materials, services of any of its discrete portions, that is designated for configuration management and treated as a single entity in the configuration management process.
NOTE	A configuration item may contain other configuration item(s).
Consequence	An outcome of an event.
NOTE 1	There can be more than one consequence from one event.
NOTE 2	Consequences can range from positive to negative. However, consequences are always negative for safety aspects.
NOTE 3	Consequences can be expressed qualitatively or quantitatively.
Correlation	The tendency for two measures or variables, such as height, or weight, or other, to vary together or be related for individuals in a group.
Cost (price)	That which must be given or surrendered to acquire, produce, accomplish or maintain something.
Data	Information represented in a manner suitable for automatic processing.
Design to minimum risk	Design of a product to an acceptable residual risk solely by compliance to specific safety requirements, other than failure tolerance.
Development	The process by which the capability to adequately implement a technology or design is established before manufacture.
NOTE	The process may include the building of various partial or complete models of the products and assessment of their performance.
Event	The occurrence of a particular set of circumstances.
NOTE 1	The event can be certain or uncertain.
NOTE 2	The event can be a single occurrence or a series of occurrences.
NOTE 3	The probability associated with the event can be estimated for a given period of time.
Factor	In measurement theory, a statistically derived, hypothetical dimension that accounts for part of the intercorrelations among tests. Strictly, the term refers to a statistical dimension defined by a factor analysis, but it is also commonly used to denote the psychological construct associated with the dimension. Single-factor tests presumably assess only one construct; multi-factor tests measure two or more constructs.

(Continued)

Box 4.3 (Continued)	
Field test	A test administration used to check the adequacy of testing procedures in the actual normal service, generally including test administration, test responding, test scoring, and test reporting.
Harm	Physical injury or damage to health of people, or damage to property or the environment.
Harmful event	Occurrence in which a hazardous situation results in harm.
Hazard	Potential source of harm. In other words, a condition, associated with the design, operation or environment of a system, that has the potential for harmful consequences.
Hazard analysis	Systematic and iterative process of identification, classification and reduction of hazards.
Hazard control	Preventive or mitigation measure, associated to a hazard scenario, which is introduced into the system design and operation to avoid the events.
Hazard reduction	Process of elimination or minimization and control of hazards.
Hazardous situation	Circumstance in which people, property or the environment are exposed to one or more hazards.
Human error	The failure of a person to perform an action as required.
Human factors in general	Is the scientific discipline concerned with the understanding of the interactions between humans and other elements of a system.
Human factors engineering	Is the scientific discipline dedicated to improving the human–machine interface and human performance through the application of knowledge of human capabilities, strengths, weaknesses, and characteristics.
Information	Intelligence or knowledge capable of being represented in forms suitable for communication, storage or processing.
NOTE	Information may be represented, for example, by signs, symbols, pictures, or sounds.
Item	Anything which can be individually described and considered.
NOTE	An item may be, for example: (a) an activity or process or product; an organization, system or person, or (b) any combination thereof.
Life cycle	Consists of three basic phases: research and development, production or construction, operation and maintenance.

Maintainability prediction	An activity performed with the intention of forecasting the numerical values of a maintainability performance measure of an item, taking into account the maintainability performance and reliability performance measures of its subsystems, under given operational and maintenance conditions.
Model	A physical or abstract representation of relevant aspects of an item or process that is put forward as a basis for calculations, prediction, or further assessment; to create or use such a model.
Multi-environmental complex of field input influences	Consists of temperature, humidity, pollution, radiation, wind, snow, fluctuation, and rain. Some basic input influences combine to form a multifaceted complex. For example, chemical pollution and mechanical pollution combine in the pollution complex. Most of these interdependent factors are interconnected, and interact simultaneously in combination with each other.
Normative reference	A reference which incorporates requirements from a cited publication into a normative document.
Output variables	Are the results of input influences interaction. Output variables can be loading, tension, output voltage and other. The output variables lead to degradation (deformation, crack, corrosion, vibration, overheating) and failures of the product.
Procedure	Specified way to perform an activity.
NOTE 1	In many cases, procedures are documented (e.g., quality system procedures).
NOTE 2	When a procedure is documented, the term "written procedure" or "documented procedure" is frequently used.
NOTE 3	A written or documented procedure usually contains the purposes and scope of an activity; what shall be done; what materials, equipment and documents shall be used; and how it shall be controlled and recorded.
Product assurance	A discipline devoted to the study, planning and implementation of activities intended to assure that the design, controls, methods and techniques in a project result in a satisfactory level of quality in a product.
Quality assurance	All the planned and systematic activities implemented within the quality system and demonstrated as needed, to provide adequate confidence that an entity will fulfill requirements for quality.
Quality control	Operational techniques and activities that are used to fulfill requirements for quality.

(Continued)

Box 4.3 (Continued)	
Qualitative data collection	Is the collection of softer information, for example, reasons for an event occurring.
Quantitative data collection	Is the collection of data that can be stated as a numerical value.
Reliability	Is the ability of an item to perform a required function under given conditions for a given time interval.
NOTE 1	It is generally assumed that the item is in a state to perform this required function at the beginning of the time interval.
NOTE 2	The term "reliability" is also used to denote the non-qualified ability of an item to perform a required function under conditions for a specified period of time.
Reliability critical item	An item which contains a single point failure with a failure consequence severity classified as catastrophic, critical or major.
Reliability growth	A condition characterized by a progressive improvement of a reliability performance measure of an item with time.
Reliability testing	Is testing during actual normal service use that offers initial information for the evaluation of the measurement of reliability indicators during the time of provided testing.
NOTE 1	If uses for accurate reliability prediction during any time (service life, warranty period, or other), it is identical to accelerated reliability or durability testing.
Residual Risk	The risk remaining in a system after completion of the hazard reduction and control process.
Risk	The combination of the probability of an event and its consequence or a quantitative measure of the magnitude of a potential loss and the probability of incurring that loss.
NOTE 1	The term "risk" is generally used only when there is at least the possibility of negative consequences.
Risk category	A class or type of risk (e.g., technical, legal, organizational, safety, economic, engineering, cost, schedule).
Risk criteria	The terms of reference by which the significance of risk is assessed.
NOTE	Risk criteria can include associated cost and benefits, legal and statutory requirements, socio-economic and environmental aspects, the concerns of stakeholders, priorities and other inputs to the assessment.

Risk evaluation	Procedure based on the risk analysis to determine whether the tolerable risk has been achieved.
Risk assessment	Overall process comprising a risk analysis and a risk evaluation.
Risk reduction	Implementation of measures that leads to reduction of the likelihood or severity of risk.
Risk trend	The evolution of risks throughout the life cycle of a project.
Safety critical function	A function which, if lost or degraded, or which through incorrect or inadvertent operation, could result in a catastrophic or critical hazardous event.
Safety measure	Means that eliminates a hazard or reduces a risk.
Severity of safety	A classification of a failure or undesired event according to the magnitude of its possible consequences.
Software	Programs, procedures, rules and any associated documentation pertaining to the operation of a computer system.
Stress testing	Classified as: Constant stress, step stress, cycling stress, and random stress.
Random stress	Is acceptable for accelerated reliability testing.
Systems engineering	Is a discipline concerned with the architecture, design, and integration of elements that when taken together comprise a system. Systems engineering is based on an integrated and interdisciplinary approach, where components interact with and influence each other. In addition to the technological systems, systems considered include human and organizational systems where the incorporation of critical human factors with other interacting factors directly affects achieving the enterprise objectives.
Systems of systems	Are composed of components that are systems in their own right (designed separately and capable of independent action) that work together to achieve shared goals.
Compliance test	A test used to show whether or not a characteristic or a property of an item complies with the stated requirements.
Laboratory test	A compliance test or a determination test made under prescribed and controlled conditions which may or may not simulate field conditions.

(Continued)

Box 4.3 (Continued)

Endurance test	A test carried out over a time interval to investigate how the properties of an item are affected by the application of stated stresses and by their time duration or repeated application.
Test development	The process through which a test is improved, planned, constructed, evaluated, and modified, including consideration of content, format, administration, scoring, item properties, scaling, and technical quality for its intended purpose.
Test development system	A generic name for one or more programs that allow a user to author, and edit items (i.e. questions, choices, correct answer, scoring scenarios and outcomes) and maintain test definitions (i.e. how items are delivered with a test).
Testing techniques	Can be used in order to obtain a structured and efficient testing which covers the testing objectives during the different phases in the life cycle.
Validation	Confirmation by examination and provision of objective evidence that the particular requirements for a specific intended use are fulfilled.
NOTE 1	In design and development, validation concerns the process of examining a product to determine conformity with user needs.
NOTE 2	Validation is normally performed on the final product under normal operating conditions. It may be necessary at earlier stages.
Whole-life costs (WLC)	Total recourse required to assemble, equipment, sustain, operate and dispose of a specified asset as detailed in the plan at defined levels of readiness, reliability, performance, and safety.
NOTE	WLC also includes the costs to recruit, train and retain personnel as well as the costs of higher organizations.

PREPARED BY SAE SUBCOMMITTEE G-11R, RELIABILITY OF COMMITTEE G-11, RELIABILITY, MAINTAINABILITY, SUPPORTABILITY AND PROBABILISTIC METHODS

Box 4.4 Draft of SAE G-11, JA1009 Standard Reliability Testing—Strategy

	SURFACE VEHICLE/ AEROSPACE RECOMMENDED PRACTICE	**SAE** JA1009 PropDft APR2013
SAE *International*® AEROSPACE		Issued Proposed Second Draft 2013-04-02
Reliability Testing—Strategy		

RATIONALE

SAE JA1009, Reliability Testing, is being completely revised to incorporate technical content and to better organize the material.

FOREWORD

This standard is to provide a basis for obtaining the information for accurate prediction of reliability, quality, safety, supportability, maintainability, and life cycle cost in real-world conditions during any part of the life cycle (warranty period, service life, and other), and to reduce the possibility of conflicts, duplications, and incorrect interpretations either impressed or implied elsewhere in the literature.

The following criteria were used to determine whether to include content in this document.

1. Reliability testing is identical to accelerated reliability testing if reliability testing uses for accurate reliability and durability prediction during service life, warranty period, or other time. Therefore, below is used often the term accelerated reliability testing instead of term reliability testing.
2. The purpose of this SAE Standard is to standardize the strategy that helps to obtain information for accurate prediction, during design and manufacturing, of reliability, durability, and maintainability in the real-world conditions.
3. The real-world conditions consist of full input influences integrated with safety and human factors.

SCOPE

This document describes reliability testing that is performed to support aerospace applications.

(Continued)

Box 4.4 (Continued)

Purpose
The purpose of this standard is to define the strategy of Accelerated Reliability Testing to give a common meaning of this strategy for contractors and users in aircraft, aerospace, and other areas.

REFERENCE

Applicable Documents
The following publications form a part of this document to the extent specified herein. The latest issue of SAE publications shall apply. The applicable issue of other publications shall be the issue in effect on the date of the purchase order. In the event of conflict between the text of this document and references cited herein, the text of the document takes precedence. Nothing in this document, however, supersedes applicable laws and regulations unless a specific exemption has been obtained.

SAE Publications
Available from SAE International, 400 Commonwealth Drive, Warrendale, PA 15096-0001, Tel: 877-606-7323 (inside USA and Canada) or 724-776-4970 (outside USA), www.sae.org
   SAE International Reliability Testing Standard JA 1009/1

ASTM Publications
Available from ASTM International, 100 Barr Harbor Drive, P.O. Box C700, West Conshohocken, PA 19428-2959, Tel: 610-832-9585, www.astm.org
   ASTM E2696–09 Standard Practices for Life and Reliability Testing Based on the Exponential Distribution

ECSS Publications
Available from European Cooperation for Space Standardization (ECSS). ECSS Secretariat, ESA ESTEC, P.O. Box 299, 2200 AG Noordwijk, The Netherlands, Phone: +31-71-565 5748, Fax: +31-71-565 6839, ecss-secretariat@esa.int

ECSS-Q-30B	Space Product Assurance. Dependability.
ECSS-M-00-03A	Space Project Management. Risk Management.
ECSS-Q-40-02A	Space Product Assurance. Hazard Analysis.
ECSS-Q-40B	Space Product Assurance, Safety.

IEC Publications
Available from International Electrotechnical Commission, 3, rue de Varembe, P.O. Box 131, 1211 Geneva 20, Switzerland, Tel: +44-22-919-02-11, www.iec.ch.

CEI IEC 61508-1	Functional Safety of Electrical/Electronic/Programmable Electronic Safety-Related Systems. Part 1: General Requirements
IEC 60300–3–2, Ed.2	Dependability Management—Part 3-2: Application Guide—Collection of Dependability Data from the Field
IEC 60605-2	Reliability of Systems, Equipment, and Components—Part 10: Guide to Reliability Testing Section 10.2 Design of Test Cycles

ISO Publications

Available from American National Standards Institute, 25 West 43rd Street, New York, NY 10036-8002, Tel: 212-642-4900, http://www.ansi.org/

| ISO 9000:2000 | Quality Management Systems—Fundamentals and Vocabulary. |
| ISO 14121 | Safety of Machinery—Principles of Risk Assessment |

Joint IEC/ISO Publications

| IEC/ISO Guide 51 | Safety Aspects—Guidelines for their Inclusion in Standards |
| ISO/IEC JTC1/SC7 | Software engineerin—Software Life Cycle Processes—Risk Management |

US Government Publications

Governmental Publications

Available from the Document Automation and Production Service (DAPS), Building 4/D, 700 Robbins Avenue, Philadelphia, PA 19111-5094, Tel: 215-697-6257, http://assist.daps.dla.mil/quicksearch/.

MIL-HDBK-108	Sampling Procedures and Tables for Life and Reliability Testing (Based on Exponential Distribution)
MIL-HDBK-217F	Reliability Prediction of Electronic Equipment
MIL-STD-690D	Failure Rate Sampling Plans and Procedures
MIL-STD-756B	Reliability Modeling and Prediction
MIL-HDBK-781	Reliability Test Methods, Plans and Environments for Engineering Development, Qualification, and Production.
MIL-HDBK-781A	Handbook for Reliability Test Methods, Plans, and Environments, Engineering, Development, Qualification, and Production
MIL-STD-781D	Reliability Testing for Engineering Development, Qualification and Production
MIL-STD-882C	System Safety Program Requirements
MIL-STD-2074	Failure Classification for Reliability Testing
DoD 3235.1H	Test and Evaluation of Systems Reliability, Availability, and Maintainability: A New Primer

(Continued)

Box 4.4 (Continued)

Other Publications

Chan, H. Antony, T. Paul Parker, Charles Felkin, Antony Oates, 2000, *Accelerated Stress Testing*. IEEE Press.

Lev Klyatis, 2016, *Successful Prediction of Product Performance. Quality, Reliability, Durability, Safety, Maintainability, Life Cycle Cost, Profit, and Other Components*. SAE International.

Klyatis Lev M., 2012, *Accelerated Reliability and Durability Testing Technology*, John Wiley & Sons, Inc.

Klyatis, Lev M., Eugene L. Klyatis, 2006, *Accelerated Quality and Reliability Solutions*, Elsevier, UK.

Nelson, Wayne, 1990, *Accelerated Testing*, John Wiley & Sons, New York, NY.

Reliability Toolkit: Commercial Practices Edition. Reliability Analysis Center. 1993.

STRATEGY OF ACCELERATED RELIABILITY TESTING

General Components of Reliability Testing Strategy
General components of reliability testing strategy are shown in Figure 4.14a

The study of real world data collection-conditions to determine the important parameters that are to be simulated in the laboratory

Figure 4.14a General components of reliability testing technology. SAE International Reliability Testing Standard JA1009/1.

Development an accurate (by quality and quantity) simulation of the real world conditions on the actual product

Introduce ART/ADT, including the analysis management and operator's influence on the causes of degradation (failures)

Make an accurate prediction of reliability, durability, maintainability, and life-cycle costs

Accelerate development of the product reliability, safety, and life cycle cost

SPECIFIC COMPONENTS OF RELIABILITY TESTING STRATEGY

One needs to study real world data connection—conditions to determine the important parameters that are to be simulated in the laboratory

Quantitative and Qualitative Data Collection
Quantitative data collection is the collection of data that can be stated as a numerical value. Qualitative data collection is the collection of softer information, for example, reasons for an event occurring. Both data types are important and support each other. The type of data collected will depend on the sort of questions to be answered by the data.

Contents and Methods of Data Collection
Data collection is providing for accurate simulation of real world conditions for reliability testing. These conditions include full field input influences, output variables, safety and human factors, reliability, durability, and maintenance data. The conditions are collected over several years and involves many different users and maintenance personnel. Thus data collection is a large-scale effort with possible sources of data corruption. Accordingly, the data collection, collation and recording process has to emphasize ease of use and error proofing.

Reliability, durability, and maintenance data are the result of field conditions action on the product and share many common elements. Therefore, reliability data collection should be integrated with the maintenance record system. Whenever possible, all data sharing of common elements should be integrated with the maintenance record system. The accuracy of data reporting can be increased if the reporting form is combined with other types of reporting, for example, economic compensation (spare costs, payment under guarantee, mileage compensation, and time reporting for the repair person). The quality of the reporting improves if the repair person knows how the data will be used. Additionally, they should be notified if their data reporting is incomplete or ambiguous.

Data collection can be automated or semi-automated by using electronic data-logging devices. The most complex data collection uses built-in electronics to perform the same task.

Automated Data Collection (ADC), also known as Automated Identification (AutoID) and Automated Identification and Data Capture (AIDC) (incorrectly referred to as "bar coding" by many) consists of many technologies including some that have nothing to do with bar codes.

Field Conditions
There are three integrated complexes present in the field conditions:

- full complex of real world input influences;

(Continued)

Box 4.4 (Continued)

- safety problems;
- human factors.

First, the full complex of real world input influences, as well as two others, are common for many types or similar groups of products and processes. The details are more specific for each application. For most of them, the full field input influences consist of multi-environmental, mechanical, electrical and electronic groups. Each group is also a complex of sub components. A multi-environmental complex of field input influences consists of temperature, humidity, pollution, radiation, wind, snow, fluctuation, air pressure, and rain. Some basic input influences combine to form a multifaceted complex. For example, chemical pollution and mechanical pollution combine in the pollution complex. The mechanical group of input influences consists of different less complicated components. The specifics of this group of influences depend upon the product or process details and functions. The electrical group of input influences also consists of several different types of simpler influences such as input voltage, electro-static discharge, and others. These factors are interdependent and interconnections, and interact simultaneously in combination with each other. Accurate simulation needs simulation of the above interconnection.

Safety problems are the combination of two basic components: risk problems and hazard analysis. Both are connected with reliability. For example, failure tolerance is one of the basic safety requirements that are used to control hazards. Another example is safety risks that are the result of the hazardous effects of failure functions. Fault tree analysis may be used to establish a systematic link between the system-level hazard and the contributing hazardous event at the subsystem, equipment or piece-part level.

The same logic applies to the solution of risk problems. There are many standards in reliability and safety that include the interconnection of reliability with safety.

Each of two basic safety components is a combination of subcomponents. The solution to the risk problem is found in the following subcomponents:

- risk assessment
- risk management
- risk evaluation.

Each of the above subcomponents consists of sub-subcomponents.

Solving the safety problem requires the simultaneous study and evaluation of the full complex of these interacting components and subcomponents.

To obtain information for risk assessment one needs to know the

- limits of the machinery
- accident and incident history

- requirements for the life phases of the machinery
- basic design drawings that demonstrate the nature of the machinery
- statements about damage to health.

For risk analysis one needs:

- identification of hazards
- methods of setting limits for the machinery
- risk estimation.

Human factors engineering is the scientific discipline dedicated to improving the human–machine interface and human performance through the application of the knowledge of human capabilities, strengths, weaknesses, and characteristics.

Human factors always interact with reliability and safety because the reliability of the product has a connection with the operator's reliability and capability. It is called human factors in the USA, whereas in Europe, it is often called ergonomics.

This is an umbrella term for several areas of research that include

- human performance
- technology
- human–machine interaction.

The term used to describe the interaction between individuals and the facilities and equipment, and the management systems is human factors. The discipline of human factors seeks to optimize the relationship between technology and humans. The human factors apply information about human characteristics, limitations, perceptions, abilities, and behavior to the design and improvement of objects and facilities used by people. The essential goal of human factors is to analyze how people are likely to utilize a product or process. Then to design it in such a way that its use will feel intuitive to them facilitating successful operation. Human factors practitioners come from a variety of backgrounds. They are predominantly psychologists (cognitive, perceptual, and experimental) and engineers. Designers (industrial, interaction, and graphic), anthropologists, technical communication specialists and computer scientists also contribute to the field.

Areas of interest for human factors practitioners usually include:

- workload
- fatigue
- situational awareness
- usability
- user interface
- ability to learn

(Continued)

Box 4.4 (Continued)

- attention
- vigilance
- human performance
- human reliability
- human–machine interaction
- control and display design
- stress
- data visualization
- individual differences
- aging
- accessibility
- safety
- shift work
- work in extreme environments
- human error
- decision-making

The classification scheme for human factors/ergonomics is much broader than the list above. The human characteristics include:

- psychological aspects
- anatomical aspects
- group factors
- individual differences
- psychophysiological state variables
- task-related factors.

Development of an Accurate Simulation of the Real World Conditions in the Laboratory

Simulation of the real world conditions consists of sub components. The first sub component is selecting a representative region. The concept of selecting a representative region for testing is used to extend real life analysis techniques and to establish the characteristics of the critical input (or output) processes. This leads to the concept of using a representative region that is the most characteristic of the total population representing all anticipated operating regions. Simulation is a tool that uses a representation or model during testing. Simulation is used to evaluate the potential results as a form of testing.

Directly Accurate Simulation of the Real World Conditions

There are different types of simulations: physical, interactive, computer (software), mathematical, and others. This standard considers the physical simulation of the field circumstances applicable to an actual product or process.

The standard considers physical simulation to be the simulation of the actual usage conditions representative of those experienced by the real product or process. Thus in the standard the items used in the simulation are typically not smaller or cheaper than those in the real object or system.

For accelerated reliability testing of the product the real world conditions need to be simulated in the laboratory where artificial input influences are used to model the actual field influences, as well as special field testing (Figure 4.14e). These conditions are in physical contact and interact with the test subject. The physical simulation of the field input influences has to be accurate by quality and quantity in order to provide accelerated reliability and durability testing.

The first basic problem is the need to understand what kind of field input influences will be simulated in the laboratory, and the purpose of these physical simulations influences.

Understanding how the various types of input influences needed for testing act upon the test subject in the field during its operation and storage (Figure 4.14b) is required. These influences include temperature, humidity, pollution, radiation, road features, air pressure and fluctuations, input voltage, and many others $(X_1, \ldots, X_N)$.

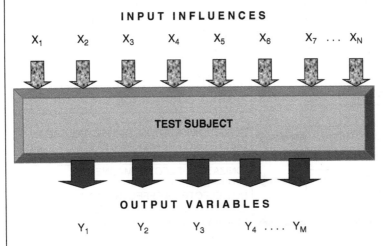

INPUT INFLUENCES

X_1 X_2 X_3 X_4 X_5 X_6 X_7 ... X_N

TEST SUBJECT

OUTPUT VARIABLES

Y_1 Y_2 Y_3 Y_4 .... Y_M

Figure 4.14b Scheme of input influences and output variables of the actual product.

The direct results of their action are output variables [vibration, loading, tension, output voltage, and many others $(Y_1, \ldots, Y_M)$]. The output parameters lead to degradation (deformation, crack, corrosion, overheating) and failures of the product.

(Continued)

Box 4.4 (Continued)

Always simulate the full range of input influences $(X_1, \ldots, X_N)$ in the laboratory when reliability testing the product.

An accurate physical simulation occurs when the physical state of output variables in the laboratory differs from those in the field by no more than the allowable limit of divergence. There are two steps in evaluating accurate physical simulation. The first step: the simulation is accurate if the output variables (vibration, loading, tensions, voltage, amplitude and frequency of vibration) during testing differs from the same output variables in the real world by no more than a given limit (for example, 3%).

This means that the output variables obey the following inequalities:

$$Y_{1\,\text{FIELD}} - Y_{1\,\text{LAB}} \leq \text{given limit (for example, 1\%, 2\%, 3\%, 5\%).}$$

$$Y_{M\,\text{FIELD}} - Y_{M\,\text{LAB}} \leq \text{given limit (1\%, 2\%, 3\%, 5\%).}$$

The second and final step involves a determination of whether the simulation is sufficiently accurate. This requires that the difference between the physics-of-degradation process during ART and in real world operations is not more than a given fixed limit.

The degradation mechanism can be estimated through its degradation parameter during product testing (Figure 4.14c). In real-world mechanical, chemical, physical, electronic, electrical, and other types of degradation mechanisms are usually interacted with each other.

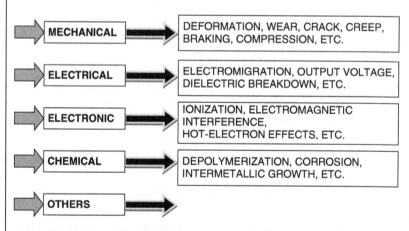

Figure 4.14c Types of physics-of-degradation mechanisms and their parameters.

Basic Components of Simulation Strategy Implementation
The following basic components of simulation strategy must be implemented to conduct ART and ADT in the laboratory to offer initial information for accurate

prediction of reliability, durability, maintainability, and life cycle cost in the real world:

- Obtain accurate data collection from the real world.
- Provide accurate simulation of the field conditions using given criteria.
- Conduct simulation testing 24 hours a day, every day, but not including
  - idle time (breaks, interruptions) or
  - time operating at minimum loading that does not contribute to failure.
- Accurately conduct simulation of each group of field input influences (multi-environmental, electrical, mechanical) in simultaneous combinations.
- Consider input influences as random processes.
- Use a complex system to model each interacting type of field influences, as well as human factors and safety.
- Simulate the whole range of each type of field influences, human factor, and safety, and their characteristics.
- Use the physics-of-degradation process as a final criterion for accurate simulation of the field conditions.
- Treat the system as interconnected using a systems of systems approach.
- Consider the interaction of components (of the test subject) within the system.
- Conduct laboratory testing, in combination with special field-testing, as components of ART/ADT.
- Reproduce the complete range of field schedules and maintenance or repair actions.
- Maintain a proper balance between real world and laboratory conditions.
- Correct the simulation systems after an analysis of the degradation and failures in the real world and during ART/ADT.

The entire range of each input influence (or variable) has to be accurately simulated. The input influences are part of a complicated process, and changing them usually produces randomness.

The accurate simulation of the input processes occurs when the statistical characteristics (such as mathematical expectation μ, variance D, normalizing correlation $\rho(\tau)$, and power spectrum) $S(\omega)$ of all the input influences, or output variables, differ from the measurements under operating conditions by no more than a specified limit, which is normally defined as a percentage.

The final conclusion about accurate simulation can be drawn after evaluating the degradation (failure) process. If the degradation or failure distribution function after ART is termed $F_a(x)$ and under operating conditions is $F_0(x)$, the measure of the difference is given as:

$$\eta[F_a(x), F_0(x)] = F_a(x) - F_0(x)$$

The function $\eta[F_a(x), F_0(x)]$ has the limit η_A (maximum of difference).

(Continued)

Box 4.4 (Continued)

When

$$\eta[F_a(x), F_0(x)] \leq \eta_A$$

it is possible to determine the reliability using the ART results. But if

$$h[F_a(x), F_0(x)] > \eta_A$$

then it is not recommended to predict reliability using the ART results.

If the values of the functions $F_a(x)$ and $F_0(x)$ are not known, it is still possible to construct graphs of the experimental data for $F_a(x)$ and $F_0(x)$ and determine the difference

$$D_{m,n} = \max[F_{ae}(x) - F_{0e}(x)]$$

where F_{ae} and F_{0e} are empirical distributions of the reliability function observed by testing the product under operating conditions using accelerated testing.

One can also find $D_{m,n}$ from graphs of $F_{ac}(x)$ and $F_{0c}(x)$. To do this it is necessary to determine the value of

$$\lambda_0 = \frac{\sqrt{mn}}{(m+n)(D_{m,n} - \eta_A)}$$

where n is the number of failures in operating conditions and m is the number of failures in the laboratory.

The correspondence between comparable distribution functions is evaluated using the probability

$$P\left[\frac{\sqrt{mn}}{(m+n)(D_{m,n} - \eta_A)} \geq \lambda_0\right] < 1 - F(\lambda_0)$$

If $[1 - F(\lambda_0)]$ is small (not more than 0.1 typically), then

$$\max[F_{ac}(x) - F_{0c}(x)] > \eta_A$$

The final choice of influences for accurate simulation is determined by the specific requirements for the test subject under operating conditions.

Each product has different parameters for the degradation mechanism. For an accurate simulation, one must evaluate the limits for each parameter.

A fundamental principle of stress testing is: more stress means greater acceleration and a lower correlation of accelerated testing results with the field results.

In real-world operation, the mechanical, chemical, and physical types of degradation mechanisms often interact with each other. The degradation parameters for mechanical degradation are deformation, wear or cracks, and other changes. The degradation parameters for chemical degradation are corrosion, depolymerization, intermetallic growth, and other changes.

The degradation parameters for electrical degradation are electromigration, electrostatic discharge, dielectric breakdown, and other changes. The degradation parameters for electronic degradation are ionization, electromagnetic interference, hot-electron effects, beta degradation, stress migration, and other changes.

If sensors can be employed to evaluate these basic parameters, then one can measure their rate of change during a particular time to compare real world and laboratory testing, and determine how similar the ART conditions are to real life conditions. If these processes are similar, then there may be a sufficient correlation between the ART/ADT results and the field-testing results.

Introduce ART/ADT, including the analysis and management of the causes of degradation

Basic Advantages of ART/ADT
ART/ADT is based on accurate simulation of the real world conditions, therefore it offers initial information for accurate prediction of the product's reliability. Studying the real world conditions in the laboratory with their influences on the test subject offers the possibility for quickly finding the right causes of failures and degradations. This provides the possibility for recommendations to accelerate development of remedies to the causes of failures and degradation. Accelerated development during the design process and improvements during the manufacturing process of the basic quality and reliability indexes of the product during any time (warranty period, service life, etc.) are possible as a result of making an accurate prediction of reliability, durability, maintainability, and life cycle costs during any given time.

ART/ADT Technology
ART/ADT performance is the technology provided for this approach to testing. Technology consists of two basic components: methodology and equipment.

The ART/ADT Methodology
The ART/ADT methodology has to provide a combination of accelerated laboratory testing and special field-testing (Figure 4.14d).

It uses special field tests for the evaluation of field influences that are impossible or very expensive to simulate in the laboratory. The selection of appropriate methodologies of accelerated laboratory testing and special field-testing depends on the specifics of the test subject and its use.

It is necessary to provide special field-testing for the evaluation/prediction of the stability of a product/technology and how the management and operator's reliability influence the product's reliability (Figure 4.14e).

(Continued)

Box 4.4 (Continued)

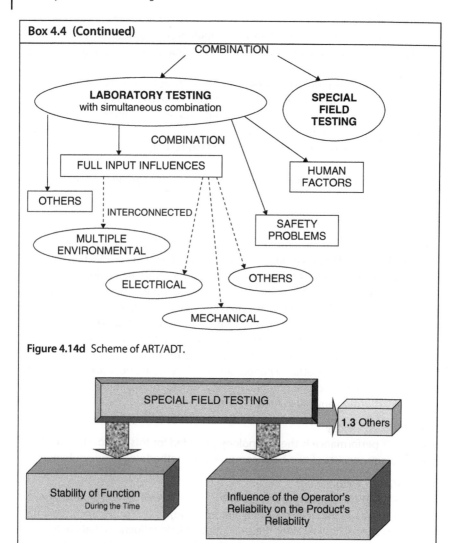

Figure 4.14d Scheme of ART/ADT.

Figure 4.14e Scheme of special field testing.

There are many different types of aircraft and aerospace products, each of which have a specific purpose and operating regime. To accurately simulate each of these it is necessary to duplicate the actual field situation to that experienced in the laboratory.

And, the operator's reliability in properly utilizing the product's is a key reliability influence that must be included in the reliability simulation. Figure 4.14f demonstrates that the operator's reliability is a function of numerous factors.

Figure 4.14f The factors that influence the product's reliability, safety, and quality through operator's and management reliability and quality.

Reliability Prediction

An important parameter in performing accurate reliability prediction is the duration of the operation, or the duty cycle for the life of the machine. This can be a long-term or short-term prediction.

- Short-term predictions assess the reliability for executing specific tasks during the workday, for example, pre-flight tests by pilots or pre-trip inspections by vehicle drivers
- As the name suggests, long-term prediction focuses on longer time segments (several days, weeks, months, years) to predict if applicants are capable of working the required period, and the changes that occur from the longer duty cycle.. For example, use long-term prediction to assess and resolve the problems of psychophysiology (PP) orientation and selection as one of the basic components of human factors.

Setting realistic limits on the levels of the abilities for operators to perform specific tasks, including functional conditions is also a concern. An example is setting limits on the complex characteristics of the management, operator's functions, and personal qualities that impact the quality of the operator's actions in controlling the vehicle over a given time period.

(Continued)

Box 4.4 (Continued)

Strategy Development of Equipment for ART/ADT

Once there is an understanding of the methodology, the second component in the development of ART/ADT technology is the beginning steps in the design of the equipment needed for employing ART/ADT functionality. First step is merging the various influences into a single piece of equipment. This necessarily will result in the combined equipment being more complicated than discrete devices. Combined equipment can consists of a combination of current types of equipment in one interconnected complex device. While this test equipment will be more complicated and expensive, it will produce more accurate real-world simulation during the testing, which results in less expense during the following steps of design, manufacturing, and product life.

Therefore, the types of combined testing equipment currently available in the market are prime candidates for ART/ADT use.

Successful development of appropriate ART/ADT technology requires a multidisciplinary team to engineer and manage the application of this technology to a particular product. As a minimum, the team should include the following:

- A team leader who is a high-level manager who understands the strategy of providing this technology, the principles of accurate simulation of the field situation, and knows what professional disciplines need to be included on the team
- A program manager who has an understanding of the design and technology needed for ART/ADT to guide the team through the process and remove any barriers that prevent the team from succeeding.
- Engineering technical staff resources
  - To perform unit filtering (selection and elimination)
  - To conduct failure analysis
  - To solve chemical problems in simulation
  - To solve physical problems in simulation
  - To predict methodology
  - To guide the system of control development, design, diagnostic and corrective action for mechanical, electrical, structural and hydraulic problems
  - To determine the need for hardware and software development and implementation
- Human factors
  - To identify and solve human factors problems when simulating them.

The team must work in close contact with departments responsible for design, manufacturing, marketing, and sales.

PREPARED BY SAE SUBCOMMITTEE G-11R, RELIABILITY OF COMMITTEE G-11, RELIABILITY, MAINTAINABILITY, SUPPORTABILITY AND PROBABILISTIC METHODS

4.2.3 Development and Implementation of Reliability Testing during the Work for the International Electrotechnical Commission (IEC), USA Representative for International Organization for Standardization (ISO), Reliability and Risk (IEC/ISO Joint Study Group)

Following presentation at the Annual Quality (ASQ) Congress, around 2001, L. K. received an invitation from Mr. John Miller, the Chair of US Technical Advisory Group for Technical Committee 56, International Electrotechnical Commission (IEC) for international standardization, to be included in the USA Technical Advisory Group—the group of experts representing the USA for IEC. Accepting this invitation, L. K. was included in the Technical Committee TC-56 Reliability and Maintainability. Figures 4.15–4.19, and 4.20 show L. K. in this committee work and work in IEC international conference; the Technical Committee TC 56 Dependability covers this area for all IEC committees in specific technical areas of energy, electronic, electrical, and so on.

The following is one story from the IEC international conference in Beijing (China). A group of this meeting's members visited the Chinese National Institute, which was involved in the testing of electrotechnical and electronic devices. During this meeting, L. K. received an award from China, as can be seen in Figure 4.18.

During this meeting and analyzing work of this institute, L. K. saw the opportunity for improvement in their reliability testing of electronic and

Figure 4.15 Meeting TC-56. Lev Klyatis is second (right) from chair.

United States National Committee of the International Electrotechnical Commission

A Committee of the American National Standards Institute

25 West 43rd Street 4ᵗʰ Fl. • New York, NY 10036 • (212)642-4936

FAX: (212)730-1346
(212)302-1286 (Sales Only)

11 October 2002

Mr. Lev Klyatis
Eccol Incorporated
72 Montgomery Street
Jersey City, New Jersey 07302

Subject: Delegate Accreditation Letter

Dear Mr. Klyatis:

The U.S. National Committee of the IEC is pleased to confirm your appointment to the USNC delegation for the announced IEC/TC 56 meeting. An Accreditation and Identification Card is enclosed for your use. Also, a USNC/IEC logo pin will be mailed to you shortly.

As you know, at this meeting you will represent the USNC/IEC. The positions on the technical agenda items will have been determined by the US Technical Advisory Group. Positions on polic and administrative matters should be developed in consultation with the USNC office. This will assure a unified US position at all levels of the IEC organization. For your information also please find the website link to the booklet "Guide for U.S. Delegates to IEC/ISO Meetings." http://web.ansi.org/public/library/intl_act/default.htm

The USNC is grateful to you and your employer for agreeing to support the voluntary standards system and for your willingness to present US positions to IEC.

Sincerely,

Charles T. Zegers

Charles T. Zegers
General Secretary, USNC/IEC

CTZ:dn

Copy to: J.A. Miller
            N.H. Criscimagna
            E.M. Yandek
            P.Kopp Ghanam

Figure 4.16 The letter from General Secretary of United States National Committee for IEC about accreditation Lev Klyatis as US Representative for IEC.

electrical devices, especially in the area of practical reliability and durability testing technology.

The first draft of this new IEC standard, "Equipment Reliability Testing. Accelerated Testing of Actual Product," was prepared, but during this time Mr. John Miller retired as chair of US Technical Advisory group and this work was stopped. Later, the ideas and information of this draft standard were used as the basis for preparing a group of six SAE International standards under the common title of "Reliability Testing."

Figure 4.17 Lev Klyatis expert of the USA Technical Advisory Group for the IEC in Sydney (Australia) during the meeting.

Unfortunately, because US representatives did not receive compensation for meeting participation, and owing to schedule and workload conflict, IEC standards were not always advanced in a timely manner. These issues are also common problems relating to the advancement of other US and international standards.

A major part of the IEC Technical Committees' work was considering proposals from countries and for accepting them as internationally recognized standards. These committees were also responsible for the development of new standards.

There are other organizations in the world that are developing international standardization. For example, the ISO, which was organized 40 years after the IEC, is one such widely known international standards organization. These organizations are planning and developing standards for all types of equipment

Figure 4.18 Dr. Lev Klyatis receiving an award from China (Beijing) during the IEC Congress.

ACOS/JSG-ISO/Klyatis/4

ACOS/JSG-ISO/Klyatis/4
2004-09-08

INTERNATIONAL ELECTROTECHNICAL COMMISSION

ADVISORY COMMITTEE ON SAFETY (ACOS)
Joint Study Group with ISO – Safety Aspects of Risk Assessment

SUBJECT
Summary of publications on risk assessment

BACKGROUND
A summary prepared by L. Klyatis of publications on risk assessment presented with the draft agenda

ACTION
For information and discussion

Figure 4.19 These documents validate Lev Klyatis as an Expert of ISO/IEC Joint Study Group in Safety Aspects of Risk Assessment.

ACOS/JSG-ISO/Klyatis/4

SAFETY ASPECTS OF RISK ASSESSMENT

September 8, Frankfurt am Main

SUMMARY

1. Assessment of Current Publications in the Area

It was reviewed Standards of 3 international organizations were reviewed: IEC, ISO, and ECSS (European Cooperation for Space Standardization).

A.*ISO – 1 standard*: ISO 14121 "Safety of Machinery. Principles of Risk Assessment". This was prepared by the European Committee for Standardization (SEN) (as EN 1050:1996) and was adopted by Technical Committee ISO/TC 199 "Safety of Machinery". This consists of 18 pages including title page, foreword, introduction, contents, scope, normative references, terms & definitions, bibliography, one empty page. So, the contents of this standard cover 13 pages, including 6 pages of Annex A (informative) and Annex B (informative).

Assessment results:
1. This standard is for machinery only which is one of many aspects of safety risk. The biological, medical, chemical, food and other aspects are not considered.
2. It includes a small (but not main) fraction of methods for analyzing hazards and estimating risk.
3. Similarly the title directions sometimes do not correspond to IEC and ECSS standards. For example, in the Simulation area in this standard only "...B.5 Fault simulation for control system" which is not essential for safety risk assessment. No less important is the "simulation of dynamic of input influences during service life", "simulation of stress conditions (vibration, environmental conditions, operator's reliability)", and others that are in IEC standards. In ECSS standards there are "dependability analysis", etc.
4. No traffic-crash aspects of risk safety are included.
5. There is no coordination with standards of other standartization organizations.
6. Other possibility may occur.

B. *ECSS* - Assessed 7 standards: 1.Glossary of Terms, 2.Space Project Management (Risk management), Space Product Assurance (3.Safety, 4.Hazard Analysis, 5.Quality Assurance for Test Centers, 6.Dependability, 7. Software Product Assurance).

Assessment results:
1. Do not take into account current 75 IEC standards in Basic Safety and 57 IEC standards in Dependability.
2. The standard in Dependability (in comparison with 57 IEC standards in Dependability) consists of nothing about techniques of dependability and no examples.
3. In standard Quality Assurance for Test Centers there are no examples.
4. Simularly by the title directions do not correspond to the contents to IEC and ISO standards. See above example (A3).
5. Standards are not coordinated with standards of other standardization organizations

(IEC).

2/3

Figure 4.19 (*Continued*)

worldwide. For coordination of the methodological aspects of this interest in advancing standardization, both organizations have joint study groups in different areas. L. K. was a member of one such joint study group, the IEC/ISO Joint Study Group in Safety Aspects of Risk Assessment, as IEC representative in this group. Figures 4.19 and 4.20 show documents from this group meeting in Frankfurt am Main (Germany), in 2004.

ACOS/JSG-ISO/Sec/7

2004-10-01

INTERNATIONAL ELECTROTECHNICAL COMMISSION

ADVISORY COMMITTEE ON SAFETY (ACOS)
Joint Study Group with ISO – Safety Aspects of Risk Assessment

SUBJECT

Report of the Joint Study Group with ISO meeting held in Frankfurt, Germany on 2004-09-08

1. Participation

Mr. F. Harless	JSG Convenor, IEC TC 44
Mr. G. Alstead	IEC TC 56
Mr. R. Bell	IEC SC 65A
Mr. N. Bischof	IEC TC 62 and SC 62B
Mr. E. Courtin	IEC TC 62
Mr. V. Gasse	IEC TC 108
Mr. H. Huhle	ACOS
Mr. L. Kylatis	IEC TC 56
Mr. I. Rolle	DKE
Mr. S. Rudnik	IEC TC 44
Mr. Y. Sato	IEC TC 56 and SC 65A
Mr. H. von Krosigk	IEC SC 65A
Mr. D. Cloutier	ISO TC 199
Mr. R. David	ISO TC 199
Ms. N. Stacey	ISO TC 199
Mr. M. Casson	Secretary, IEC CO

2. Opening of the meeting – introduction of the delegates
Mr. Harless, the JSG Convenor opened the meeting stressing that it was a Joint Study Group with ISO participation. He noted that there were three delegates from ISO/ TC 199 and that Mr. Courtin also represented ISO/ TC 210. The Secretary was requested to ensure that the report of the meeting is circulated within ISO.

The delegates then introduced themselves giving brief information on their professional activities and their IEC/ ISO affiliations.

Mr. Rolle welcomed the delegates to DKE.

3. Approval of the agenda
Mr. Harless proposed to kick-off the meeting with a presentation and then a round-table discussion should start prior to the lunch break.

The agenda was approved.

4. Presentation – Mr. Harless
Document: ACOS/JSG-ISO/Harless/3

Figure 4.20 Documents demonstrating Lev Klyatis as an Expert of ISO/IEC Joint Study Group in Safety Aspects of Risk Assessment.

Dr. Klyatis saw the need for improving the level of testing electronic and electrical devices, especially in practical reliability and durability testing, and why the implementation of his ideas in IEC and ISO standards was so important.

4.3   Implementing Reliability Testing and Prediction through Presentations, Publications, Networking as Chat with the Experts, Boards, Seminars, Workshops/Symposiums Over the World

The basic challenge of implementing any new approaches is the new technology must, first, be embraced and accepted in the minds of the people who will be doing the implementation—whether this implementation be engineering new methods or devices in research, design, or specific industrial company applications. Only when the professionals who will be implementing the new approaches fully understand and are committed to the new technology will successful implementation be possible. The author's strategy, approaches, and innovations to successful reliability prediction and ART/ADT (see Chapters 2 and 3) were learned from his publications, and then implemented in different areas of engineering, physical, mathematical, and other sciences.

An important aspect of successful implementation of new ideas and technologies is not only how new or high quality is the idea or technology and its benefits, but how expensive it will be to implement and use the new technology. This requires not just the direct costs of this technology in comparison with the present technology, but how its implementation will influence the costs (or savings) of all the subsequent processes throughout the product's life cycle.

Consider the following example of successful ART/ADT. For this implementation it was important to do a cost versus benefit analysis for the proposed ART. The question was what is the cost versus benefits of a simultaneous combination of different types of input influences, compared with single testing with separate simulations of each input influence.

In this case, the cost of the equipment, the methodology, and conducting each of the single tests (e.g., separate vibration testing, or corrosion testing, or temperature plus humidity testing, or testing in dust chambers, or input voltage plus vibration testing) may appear to be less costly than reliability testing using the simultaneous combination of simulation of all the aforementioned types of influences.

While this may seem correct, it only compares the direct cost of the two approaches to testing, and therefore is not correct. The quality of the testing level influences the cost of many subsequent processes, including design, manufacturing, use, warranty and recall, and service life. For example, the confidence level of different levels of testing impacts returns (recalls, etc.), future changes in safety, quality, and reliability problems during the product's life cycle. This is a case of where not accurately simulating the interacted real-world influences would lead to increased overall expenses, and decreases the economic profitability of the product.

It is Dr. Klyatis experience from consulting with many companies that when the test engineer working with vibration testing was asked what they were evaluating, the usual answer was "Reliability." There would then follow an explanation of why this answer is not truly correct. The product's reliability depends not only on vibration. Vibration is only one component of the many mechanical testing (influences) the product experienced in real-world use. In the real world, mechanical influences act in combination with multi-environmental influences, electrical and electronic influences, and so on. The product's reliability (the degradation process, time to failures, and mean time between failures, etc.), is a final result of all field input influences and actions, as well as human factor influences. If one does not take all of these into account, one cannot successfully predict the product's reliability, or other components of performance.

As a result of testing with inaccurate field condition simulation, during the use of the product unpredicted accidents, product failures, and other faults that were not identified in reliability testing may result in increased and unanticipated costs during usage. The incremental costs incurred through returns and the cost of improving the design and manufacturing processes are often overlooked when estimating the cost attributed to testing.

As was previously shown in the Preface of this book, over 23 years, recalls alone in the automotive industry have increased dramatically, leading to industry losses amounting to billions of dollars. For these reasons, ART (or ADT) is actually more economical than independent single-influence testing. While the testing may be more expensive, it yields decreasing costs during the manufacturing and the product's in-use life cycle. While this is true, unfortunately too often it is difficult to quantify these post-testing savings. Detailed consideration of this problem with examples is presented elsewhere [1, 4, 5].

As already mentioned, L.K.'s papers with the essence of his ART/ADT development were discussed during his lectures (see example in Figure 4.21) and also discussed and published (AGRI/MECH/43) by the United Nations Economic Commission For Europe in 1968 and 1969. The United Nations also published the final paper "Accelerated testing of agricultural machinery" in New York in 1970 (see in Figure 4.22 title page and one page from Chapter 6).

In 1989 the government of the Soviet Union first organized an engineering center, based on this work, and then organized the state enterprise Testmash for widespread implementation in other companies in the country. These utilized L. K. ideas, research results, and innovations in reliability testing and successful reliability prediction.

Lev Klyatis was also the Soviet Union representative to the US–USSR Trade and Economic Council, which was organized by the USA and the USSR governments for improving collaboration between the two countries. At the meeting of this council in New York in 1989, there were two presentations from each country. L. K. made one of these, a presentation about Testmash's solutions for development and implementation ART/ADT and reliability prediction.

Figure 4.21 Dr. Lev Klyatis during his lecture for professionals in reliability testing and prediction (Latvia, 1974).

Following this presentation, at the meeting of the US–USSR Trade and Economic Council, many executives of American companies met with Lev Klyatis. They were anxious to discuss his approaches, as they were looking for development in reliability testing and prediction using modern technologies. Two companies that were especially persistent were representatives from Steptoe and Johnson who were interested in implementing the Testmash approach into American companies.

After this presentation in New York, L. K. received a proposal from the Kamaz, Inc. (Russia) chairman Mr. N. Bekh for implementation of Testmash's system for development at Kamaz, Inc., including the system of research, design, manufacturing, and maintenance. Kamaz was one of the largest companies in the USSR's automotive industry, designing and manufacturing trucks. Employing over 100,000 employees at that time, Kamaz, Inc. sold its products mainly in the USSR and some eastern European and other countries. But KAMAZ, Inc. wanted to begin selling their products in the more developed West European, American, Asian, and other markets. As a part of this venture into the Western market, they reached an agreement that Cummins (USA) would build a special plant for them in the USSR for manufacturing Cummins' engines. They planned to use Testmash's technology for increasing product reliability, safety, durability, maintainability, profit, decreasing life-cycle cost, and to solve other problems that would open the door to the Western market through increasing the competitiveness of their product in the world market.

AGRI/MECH/43

ECONOMIC COMMISSION FOR EUROPE

AGRICULTURAL MECHANIZATION

ACCELERATED TESTING OF AGRICULTURAL MACHINERY

A. Kononenko and L. Klyatis
(USSR)

UNITED NATIONS
New York, 1970

Figure 4.22 Cover page and first page of Chapter 6 from Dr. Lev Klyatis's paper for the United Nations.

6. Physical modelling of the conditions of operation of agricultural machines,
    reproducing the complete range of operational stress schedules

In the accelerated testing of agricultural machines it i. not always possible
make a sufficiently accurate evaluation of their reliability or of the change whic
takes place in their economic performance from beginning to end of their service]

This is because the accelerated testing technique described above reproduces
severest stress schedule to be met with under operating conditions; the service li
of components which fail and the costs incurred in restoring machines to working (
differ from those found in field operation. Attempts made to devise reliable and
sufficiently accurate factors for converting the results of accelerated tests to. 1
figures which would be obtained in normal field operation have as yet met with no
success.

However, there is another possible method of accelerated testing, based on tl
principle of reproducing the complete range of operating schedules and maintaining
proportion of heavy to light loads.

This method has the following potential advantages:

One net hour of work performed by the machine with a faithfully reproduced si
schedule is identical in destructive effect with one net hour of work done ui
normal operating conditions;

Because of this, there is no need to force the pace of testing in terms of tl
size and proportion of stresses, and consequently no need to work out
acceleration factors;

The technical and economic performance figures needed to evaluate the machine
be determined at any time in the course of operation, without needing conver:

Despite the complexity of this method of accelerated testing, the present st:
of the art renders it feasible in principle.

Research into such a technique has been in progress in the USSR for some yea1
A theory for the construction of physical models simulating the working conditions
agricultural machines is being evolved, and rigs and climate rooms are being desij
for the appropriate tests.

The new technique is based on the following considerations. The stress sche(
to which a machine is subjected are one of the main factors characterizing its
operation. It is necessary to reproduce, not the maximum stresses alone, but the

Figure 4.22 (*Continued*)

Upon returning to Moscow, Dr. Klyatis learned from the people with the
USSR Automotive and Farm Machinery Department that Mr. N. Bekh was also
the advisor in economics for Mr. M. S. Gorbachov, the President of the USSR.

Mr. Bekh and Kamaz's general manager Mr. Paslov had an in-depth understanding of what successful prediction of product performance could mean

for the success of their company. They were ready to invest many millions of dollars in this implementation, because they read L. K. publications and understood that this investment would result in the company earning much more money. This is in stark contrast with his experiences with the management of many American and Japanese companies (as described in this book), who could not understand the in depth benefits that would be achieved from successful reliability prediction. Therefore, they were reluctant to accept the new ideas and approaches to reliability and durability testing as providing potential economic benefits for their companies.

The result was a contract between Kamaz, Inc. and Testmash which included all the requirements relating to implementation, including payment for Testmash's patents in several countries. With this contract, Testmash began to design and implement solutions as was required by Kamaz, Inc. The final result was the specific design of new equipment for the physical simulation of multi-environmental influences using a large test chamber that included, as one component, vibration equipment capable of testing complete three-axle trucks, corresponding to Dr. Klyatis's recommended methodological solutions.

Following the widespread implementation of Testmash, the all-USSR journal *Tractors and Farm Machinery* [6] published an interview with L. K. in which there was a discussion on why Dr. Lev Klyatis had been appointed by the USSR government as the chairman of this organization, how this was the new state-of-the-art research center for industry, what kind of problems this organization was designed to solve, and the role of Testmash in the development of the industry of the Soviet Union [6]. The basic ideas presented in this interview, the concepts, and the approaches developed are now actuality and can continue to be implemented well into the future. This interview also described the new strategy and methodology, the specific test equipment that was under development in Testmash, and the specification developed through Testmash in the 1990s in the Soviet Union, much of which is still in actual use and continues to advance in developed countries throughout the world. The first page of this interview (in Russian) is presented in Figure 4.23. Following presentations at several SAE World Congresses, there were meetings with attending professionals who understood how useful this new direction in successful prediction of product reliability could be. For example, Mr. Richard Rudy, senior manager in quality and testing DaimlerChrysler, said that he believed that this new approach was very interesting and could be useful for other industries, and he would be happy to give excellent recommendations or reviews for implementation of these ideas. Mr. Richard Rudy was for many years the SAE International representative in the Board of Directors Annual Reliability and Maintainability Symposium (RAMS), and Executive Committee for a group of technical sessions under the common title "Integrated Design and Manufacturing (IDM)" at the annual SAE World Congresses in Detroit.

Figure 4.23 First page of published interview with Dr. Lev Klyatis, Chairman Testmash.

Figure 4.24 First job for Professor Klyatis in the USA: fish delivery.

Professor Klyatis began consolidating his ideas, strategy, and implementation approach for reliability prediction and testing in the USA from presentations at the ASAE international meetings, RAMS, and ASQ Congresses. In order to obtain money for these trips to these meetings, he began to work delivering fish (Figure 4.24).

When Elsevier published his second English-language book *Accelerated Quality and Reliability Solutions* in the UK, Richard Rudy wrote the book review, which was published in the journal *Total Quality Management and Business Excellence* (UK) in September 2006 (Figure 4.25). When they met a few years later, during another SAE World Congress in Detroit, Richard described how, after their first meeting, he had proposed to DaimlerChrysler's management that they should invite L. K. to come in as a consultant for DaimlerChrysler, as it would be useful for the company to help improve their product reliability, and described why, but after discussion with his Vice President he was told there were no available positions or funding to retain L. K. Richard Rudy also wrote the published review to L. K. book *Accelerated Reliability and Durability Testing Technology*, which was published by Wiley in 2012.

From beginning his life in America in 1993, he quickly understood that many professionals in the fields of reliability prediction and reliability testing were not ready to learn and use the entire methodology partly presented in his book *Successful Accelerated Testing* (Mir Collection, New York, 2002). Therefore, he developed a method of implementing his solutions using a step-by-step

Total Quality Management
Vol. 17, No. 7, 959–960, September 2006

Book Review

Accelerated Quality and Reliability Solutions
Lev M. Kylatis & Eugene L. Klyatis.
Oxford, Elsevier, 2006, ISBN 0-08-044924-7, 544 pp. US$150/€136.

A new and very useful book on accelerated reliability testing is now available from Elsevier. The book focuses on the accurate depiction of field influences on a product's performance and how to simulate these in a laboratory environment. Whereas most books on accelerated testing concentrate on the mathematical aspects of analyzing the data, this book details how to develop the test to accurately reflect field usage, then how to duplicate this usage in actual laboratory testing. This information is sorely needed in industry today because of the need to design and develop high reliability components and systems in an ever-shorter time span in order to get the product to the marketplace before the competition.

Many companies today still rely on the test-analyze-test method to 'grow' reliability. There is much literature still being written on the outdated concept of reliability growth. It is still not generally recognized that a robust design, i.e. a design whose performance is insensitive to the variation in environments and usage conditions of the customer, must be developed with the initial design. Thus, there is one and only one chance to test the design in a representative, accelerated fashion to prove high reliability. Dr and Mr Klyatis's book provides valuable information on the how-to side of analyzing the inputs to identify the right factors that influence reliability, and then developing the accelerated test that duplicates those factors influence in the laboratory test.

The book is divided into five chapters. Chapter 1 details the strategy for developing an accurate physical representation of the influence of various field inputs. It reviews how to obtain accurate information from field, then selecting a representative input region to assure accurate field conditions in the test. It discusses the influence of climate on reliability, looking specifically at the influences of solar radiation, temperature, fluctuations of daily and yearly are temperatures, humidity rain, wind speed, and other atmospheric phenomena. The authors then show how each of these factors affects reliability of the products. The final section of chapter 1 details how to simulate these particular influences on the product using wherever necessary artificial media for these natural phenomena.

Chapter 2 deals with developing the specific accelerated reliability test for the products. It contains a detailed eleven-step process to develop the test, starting with collection of the field information and ending with using the results of the test for rapid, cost-effective improvements. It then discusses acceleration methods for solar radiation, chemical destruction, weathering, corrosion and vibration.

Chapter 3 show how to make useful reliability, durability, and maintainability predictions from the results of these accelerated reliability tests. The authors first review the mathematical basis for being able to predict from the results of accelerated testing. They then discuss the development of techniques to predict reliability without finding the analytic or graphical form of the failure distribution. This is followed by predictions using mathematical models with dependence between reliability and factors from the field and manufacturing. The chapter concludes with discussions on simple and multi-variate Weibull analysis, durability predictions, and predictions of optimal maintenance intervals and spare parts usage.

Chapter 4 expands from accelerated reliability testing to the use of accelerated methods for quality improvement and improvements in manufacturing. It reviews basic quality concepts, then shows how to use these in accelerated manufacturing improvement and design of manufacturing

1478-3363 Print/1478-3371 Online/06/070959-2 © 2006 Taylor & Francis
DOI: 10.1080/14783360600958120

Figure 4.25 Published review in the journal *Total Quality Management and Business Excellence*, Taylor & Francis Group, Volume 17, Number 7, September 2006, UK.

960 *Book Review*

equipment. It then shows how to find those manufacturing factors that influence product quality and how to use them to improve product design.

Chapter 5 deals with the basic concepts of safely and risk assessment. It covers estimating risk; evaluating risk; performing hazard analysis; and managing risk. The chapter concludes with an introduction to human factors.

This book is an excellent reference text on accelerated testing. The book is of great benefic of the design engineer level. It covers in detail the many aspects of designing and developing an accelerated reliability testing not heretofore found in other texts. The only thing would have made this book even more valuable would have been the addition of problems at the end of each major section so that it could be used as a college-level textbook for engineers. Still, the book is an excellent addition to the library of every design engineer seeking to improve his/her design expertise.

Richard J. Rudy
Senior Manager, Product & Process Integrity (retired)
DaimlerChrylser Corporation

Figure 4.25 *(Continued)*

approach. This approach first focused on the initial step: the need for accurate physical simulation of field conditions for accurate accelerated testing. This need is easily understood by most people.

This is why, for most of his early years in the USA, L. K. presentations focused on the need for accurate simulation of field conditions. And this starting point was important because so many industrial companies' research centers were not accurately simulating field conditions. Therefore testing results often were, and continue even today, to be different from the field (real-world) results.

Dr. Klyatis's second book, which was much more complicated, demonstrated solutions to actual reliability and quality problems. Around 2004, after making his presentation at the SAE World Congress in Detroit, which again was well attended, one attendee came to him and asked: "Do you have experience in book writing?" His answer was "Yes!" This was quickly followed by the question: "Would you be interested in preparing a book on your theories and strategy on quality and reliability for publishing by Elsevier?" Knowing Elsevier to be a large world class publisher, he immediately responded "Yes" The person who had asked was Jonathan Simpson, Commissioning Editor, Science and Technology Books, Elsevier, UK. He explained the process whereby, once he received the completed book proposal in the UK, they would send this book proposal for reviews, and if the reviews were positive they would sign an agreement with the author for publishing the book.

Elsevier received excellent reviews, and the book *Accelerated Quality and Reliability Solutions* was published by Elsevier in December 2005 [5]. The book contained 514 pages and was published in Oxford, UK. L. K. son, Eugene, co-authored the book. Eugene, who had worked as a quality manager for a major industrial company, prepared the chapter related to quality improvement and assisted in the preparation of another chapter. This book is available in many libraries throughout the world, and there are many published citations and reviews of this book. The story leading to his third book in English since having moved to America also has an interesting background. During the Questions and Answers segment for Dr. Klyatis's presentation at the SAE World Congress in Detroit, around 2008, one attendee commented: "I work in durability testing. I have looked for literature on how I can conduct this testing, but I found only papers, journal articles, and books, where all the information is about durability testing, but no information of how I can conduct durability or reliability testing. I need information on the technology for doing this testing. What can you recommend?"

L. K. replied: "You are right. Presently there is not much published on how to do this. So, soon I will prepare a new book in reliability and durability testing technology. Please, wait for my new book." While the answer was impromptu, this question and the need for a publication on this subject motivated him to write that third book. He understood that this book addressed problems that needed a solution and would be applicable worldwide. This required a publisher who could attract a worldwide audience. Years earlier, while in Manhattan, New York, he remembered seeing a building with the large letters WILEY, and for many years had dreamed that someday he would have a book that would be published by John Wiley & Sons. Therefore, he decided to send this book proposal first to Wiley to see whether they would be interested in publishing it.

While this proved to be a more difficult book to write, it provides a useful tool for this area and became very popular in industry, as can be seen through an Internet search. Such a search, however, does not include all references, but one can see that professionals are using this book [1] in many different areas of industry and science worldwide.

That book (Figure 4.26) reflected his step-by-step strategy and practical technology for providing advanced reliability testing [1], while the books prior to this were primarily descriptions of his basic work in the new direction in successful prediction of product reliability, and emphasized the basic interconnections of components—quality, durability, maintainability, supportability, safety, human factors, life-cycle cost, recalls, and so on.

One example of the practical application of these theories and approaches involved his work with Nissan. Having attended the presentation "Nissan speeds up truck bed durability testing" by Nissan's Technical Center representatives Mr. Ali Karbassian and Mr. Tetsufumi Katakami at the SAE 2008 World Congress in Detroit (this paper was later published in SAE magazine *Automotive Engineering International* in August 2009), Dr. Klyatis explained and worked with management to understand that the examples presented were a powerful demonstration of not enough interconnection between different departments of the company, and that changes were needed in corporate interactions to improve reliability and durability testing. This resulted in Nissan's middle and top management implementing Dr. Klyatis's approaches to ART and ADT. It should be noted that this situation was not unique to Nissan, but can be found in many other small, large, and middle-sized industrial not only in the automotive area, but in many other industries.

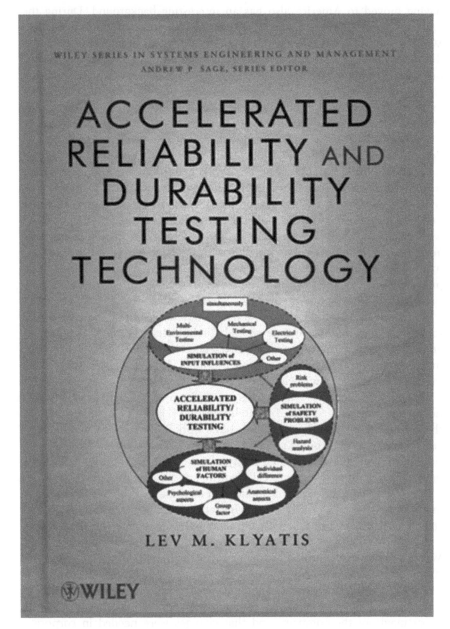

Figure 4.26 Front cover of the book [1].

A similar incident occurred during Dr. Klyatis meeting with Dan Fitzsimmons, a Boeing Fellow, working in the quality and reliability areas. After L. K. lecture for the SAE G-11 Division members in Washington, DC, Dan discussed and agreed that lack of adequate interaction between different areas of engineering in large companies is a major factor in inadequate success in reliability prediction and effectiveness.

Dr. Klyatis first three books published after arriving in America provided a basic understanding and usefulness of the components of his prediction methodology. These three books prepared them and made them ready for understanding the complete concepts and strategy of successful prediction of product performance using the interacted components – quality, safety, reliability, durability, maintainability, supportability, life cycle cost, profit, and so on. These three books provided the foundation to enable writing his fourth new book in this direction [4]. This publication needed to be published by a large engineering society, as their membership, which can be well over 100,000 interested professionals, would be the primary audience who would benefit from the book. It would also provide a marketing value by disseminating information about their publications in the society's magazines, journals, and web sites.

So, Lev Klyatis approached SAE International, who, after receiving excellent reviews of his book proposal, made some comments and recommended publishing this book. The book *Successful Prediction of Product Performance* [4] was published at the end of 2016 and provides guidance on how to implement successful prediction of product performance, including reliability, in different areas of industry. However, what L. K. did no take into account was that SAE's marketing did not directly contact society members about new publications, and so, many SAE members who would benefit from this book did not know it was available.

The RMS Partnership, a US Department of Defense (DoD) organization beginning in approximately 2000 and continuing through to today, is also involved in the implementation of reliability testing and prediction. The RMS Partnership included Dr. Klyatis in tutorials and presentations in organized meetings, and published articles in the *Journal of Reliability, Maintainability, and Systems Engineering*. They also published reviews of his books.

One example of this work involved an RMS Partnership organized in 2012 by the US DoD, the Department of Transportation, and industry for a workshop and symposium (see part of program in Figure 4.27) entitled "Road Map to Readiness at Best Cost for Improving the Reliability and Safety of Ground Vehicles." L. K. was invited as a tutor and speaker for two presentations along with several other presenters. The workshop was held in Springfield, Virginia, September 19–20, 2012. Participating organizations and companies included the US Army, NHTSA (Department of Transportation), Volpe Center, Vehicle Research; Honda Motor Company; Active Safety Engineering, Raytheon

A Road Map to Readiness at Best Cost for Improving the Reliability and Safety of Ground Vehicles

Workshop & Symposium September 19 & 20, 2012
Waterford Conference Center, Springfield, Virginia
September 19 & 20, 2012

Participating Organizations: U.S. Army, BAE Systems, NHTSA; OSD; Volpe Center, Vehicle Research, Insurance Institute for Highway Safety; Active Safety Engineering ; Honda Motor Company; Active Safety Engineering; Raytheon Integrated Defense Systems; University of Alabama, Huntsville; SoHaR, Inc.; ReliaSoft; Clockwork Solutions; Alion Science and Technology Corp., Southern Methodist University; Office of Secretary of Defense (MR) & ------more

Biographies

5:30-
6:15

Tutorial on
Reliability
Testing and
Accurate
Prediction of
Reliability &
Safety

Instructor, Dr. Lev Klyatis,
Conference Room
www.rmspartnership.org/upload/sept_rmsp_workshop.pdf

Figure 4.27 The first day's program of the DoD, Department of Transportation, and industry workshop/symposium.

Integrate Defense Systems; University of Alabama; ReliaSoft; Alion Science and Technology Corporation; Office of Secretary of Defense, and others. This workshop provided a good opportunity for discussing the implementation of the ideas, approaches, and technologies to reliability testing and prediction presented in the current book and in disseminating L. K. strategy and tactics to this audience of high-level professionals.

SAE International's Annual World Congress & Exhibition, claimed to be the largest meeting of this type in the world, is directed to all areas of mobility engineering and technology, is convened each year in the Cobo Center, in Detroit, MI. There are between eleven thousand and twelve thousand professionals from many countries in attendance each year. The attendees are primarily professionals from many different areas of mobility engineering,

A Road Map to Readiness at Best Cost for Improving the Reliability and Safety of Ground Vehicles

DAY 1 Program   September 19, 2012

8:15 – 9:00	Registration & Refreshments	Networking Opportunity
9:00 – 9:10	Conference Introduction	Dr. Russell Vacante, President, RMS Partnership
9:10 – 9:40	Opening Keynote Address (DoT)	HON. David L. Strickland, Administrator, National Highway Traffic Safety Administration (NHTSA)
9:40 – 10:10	Sustaining Readiness at Best Cost	Walter B. Massenburg, Senior Director, Mission Assurance Business Execution, Raytheon Integrated
10:10 – 10:20	BREAK	Break Area
10:20 – 11:20	Opening Keynote Address (DoD)	HON Katrina McFarland, Assistant Secretary of Defense (Acquisition), Office of the Secretary of Defense (Acquisition, Technology & Logistics)
11:20 – 12:00	Terrain Roughness and Suspension Capability - The Keys to Improved Reliability	Brett Horachek, Program Manager at Nevada Automotive Test Center
12:00 – 12:45	LUNCH	Break Area
12:45 - 2:00	Panel #1 - A Futuristic View of Ground Transportation Systems	- Melvin "Mel" Downes, Chief, Weapons, Fire Control & SW Quality, Reliability & Safety Engineering, the Quality Engineering and System Assurance Directorate, Armament Research, Development and Engineering Center, Picatinny Arsenal, NJ. - Gary Ritter, Director, Center for Advanced Transportation Technologies, RVT-90 Volpe Center - Robert E. Wild, Tactical Vehicles & Tires, Research, Development and Engineering Center, Army Tank Automotive (TARDEC), USA
2:00 – 2:30	Early Testing and Evaluation Impact on Reliability and Cost	John Paulson, Senior Director, Engineering Development and Technology, General Dynamics Land Systems
2:30 – 3:25	Panel #2 - Technology Innovation (The Impact of New Technology on Ground Vehicle Reliability, Safety & Cost)	- Joseph Dunlop, Vice President, Test Operations, Calspan Corporation - Tracy V. Sheppard, Director, Automotive Directorate, US Army Test Center, Aberdeen Providing Ground, Md. - Greg Kilchenstein, Deputy Assistant Secretary of Defense, Maintenance
3:25 – 3:35	BREAK	Break Area
3:35 – 4:05	Ground system Reliability	Dr. David Gorsich, Chief Scientist, U.S. Army Tank Automotive Research, Development and Engineering Center (TARDEC), TACOM LLC
4:05 – 5:30	Panel #3 - Reliability & Safety Standards for Ground Vehicles	- James French, Committe Manager, Mission Assurance Standards Working Group - Dr. Lev Klyatis, Senior Advisor, SoHaR, Inc. - Christopher J. Bonanti, Associate Administrator for Rulemaking, National Highway Traffic Safety Administration - Chris Peterson, Reliability Consultant, H & H Environmental Systems, Inc.
5:30 – 6:15	Tutorial on Reliability Testing and Accurate Prediction of Reliability & Safety	Instructor, Dr. Lev Klyatis, Conference Room, various conference all locations
6:15 – 7:30	Social Mixer	RMS Partnership & Sponsor Hosting

Figure 4.27 *(Continued)*

including automotive, aerospace, off-highway, farm machinery, electrical, electronic, and many others. While the attendees are primarily engineers, all levels of management from these industries can be found in attendance. Each year the Congress's papers are published by SAE International. The Event Guide for the Congress usually consists of 220–250 pages, including general information, special events and networking opportunities, ride and drive, management programs, chat with the experts, over 100 technical sessions, committee task force and board meetings, awards and recognitions, seminars,

Figure 4.28 Lev Klyatis, chairman of technical session IDM300 Trends in Development Accelerated Reliability and Durability Testing Technology, SAE 2014 World Congress, introducing a speaker from Jatko Ltd (Japan).

exhibit directory, and other important information. The 2013 SAE World Congress Event Guide contained information on Dr. Klyatis's achievements in the development of successful prediction of product reliability, including a Chat with the Experts session [7]. Since 2012 he has been a chairman and co-organizer of the technical session "IDM300 Trends in Development of Accelerated Reliability and Durability Testing," which was organized as a direct result of his work in this area (Figure 4.28).

Typically these events, World Congresses, RAMS, Quality Congresses, and similar gatherings, include tours to nearby major industrial companies. One surprising observation he made from these tours was that even very high technology large companies, such as Thermo King, Lockheed Martin, Boeing, and others, were still conducting (and continue using) separate vibration testing, temperature/humidity testing, corrosion testing with only chemical pollution of components in chambers with simulation of only a simple degradation of a single element at a time. These companies still do not understand what is necessary for successful prediction of product performance. When the Director of Quality at Thermo King Co. invited Dr. Klyatis to help the company improve their product reliability, he agreed to consult with them to improve their expertise and their equipment for durability and reliability testing. Some of this implementation work is described in greater detail elsewhere [1, 4, 5]. He observed similar situations during visits to the NASA Langley Research Center with others experts from the SAE G-11 Division in aerospace standardization, to the Lockheed Martin plant during the ASQ Congress in Denver, and many other industrial companies. In almost all of these instances he was surprised to observe that there was little implementation of process reliability and durability testing into the design and manufacturing processes. Although in the last few years he has not had the opportunity to visit these companies, from their representative presentations it is evident that the process of implementation is moving forward very slowly. For example, in the magazine *Aerospace Testing International*, which contains mostly information about flight testing, you will find articles about wind tunnel testing or vibration testing of some components, or about sensors or other testing details. One such article "The new generation in testing is here" appeared in the June 2015 edition. This article provides a description of a horizontal shaker for a vibration testing machine. But professionals know that horizontal shakers do not accurately simulate real-world vibrations of test subjects, especially for mobile machinery applications. This represents technology that has been in use for more than 100 years, and because of the dynamics of mobile machinery cannot be used for successful reliability or durability testing. Therefore, it cannot provide the information necessary for successful reliability prediction. It also must be reiterated that successful implementation of new ideas and technologies depends not only on the strength of the ideas and technology, but must be based directly on the benefits it will provide: how expensive the new technology is in comparison with present technology, and the savings it will accrue to the organization in subsequent processes.

It is also somewhat encouraging that we are seeing that in the advanced research centers over the world there is movement in the direction of new test chamber development that includes multiple variables. Canada developed one such test chamber (Figures 4.29 and 4.30), described in further detail in the author's book *Successful Prediction of Product Performance* (Figure 4.31) [4]. Detailed consideration of this problem, with examples, can be found in some of his other publications [8–10].

Dr. Lev Klyatis also attended and provided oral and written presentations over a period of five years at the Annual Reliability and Maintainability Symposiums (Figure 4.32), and at several IEEE Workshops on Accelerated Stress Testing (Figures 4.33, 4.34, and 4.35). These events included discussions with colleagues from different countries regarding the problems in testing; the new ideas for successful reliability prediction; and understanding that sometimes an organization was not ready to understand the importance and usefulness of the entirety of the concepts and strategy of ART/ADT for successful reliability prediction. Therefore, many of these presentations related only to components of the methodology for successful prediction of the product, leaving the complete development of methodology simulation, reliability testing, and prediction for the service life for another time.

One of the most popular presentations was "ESTABLISHMENT OF ACCELERATED CORROSION TESTING CONDITIONS" at the Annual RAMS (The

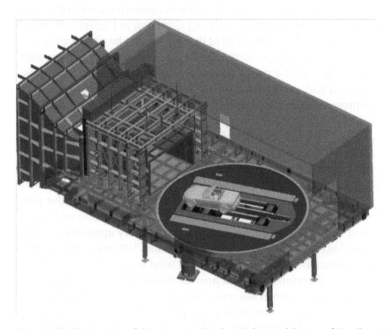

Figure 4.29 The system of drive-in test chamber (Advanced Center of Excellent, University of Toronto, Canada).

ANSWER TO INVITATION FOR SAE 2016 WORLD CONGRESS PRESENTATION

"Hi Lev,
I spoke to the Director at ACE. We are currently not working too much in reliability. This will change in the future however as we are going in this direction. We won't have anything for September 1st, but please keep us updated for future opportunities.
Thanks,

Colin, ACE Marketing Manager ", (2015)

Figure 4.30 E-mail response from ACE (Canada) to Lev Klyatis's invitation for presentation of an ACE solution at the SAE World Congress.

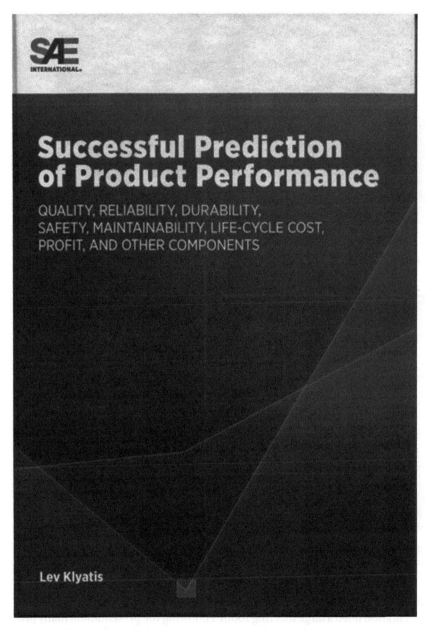

Figure 4.31 Front cover of the book published by SAE International.

Figure 4.32 Lev Klyatis, presenter at the RAMS. Paul Parker is a chair of the technical session.

Figure 4.33 Lev Klyatis, panel presenter at the IEEE Workshop on Accelerated Stress Testing, Pasadena, CA, 1998.

International Symposium of Product Quality and Integrity), January 2002. This presentation filled a large meeting room. Following the presentation, many professionals from industrial companies waited to speak with this author to discuss how they could implement this new approach to corrosion testing in their organization. It was recommended that they implement advanced corrosion test methodology for their complete mobile products such as automotive, farm

SPECIFICS OF ACCELERATED
RELIABILITY TESTING

♦ LEV KLYATIS
♦ Habilitated Dr.-Ing., Sc.D., PhD
♦ Professor Emeritus

| Lev Klyatios | ASTR 2009 Oct 7 – Oct 9, Jersey City, NJ | Abbreviated Title Page 1 July 9, 2016 |

Figure 4.34 The title page of the visuals during the presentation at the IEEE ASTR Workshop, 2009.

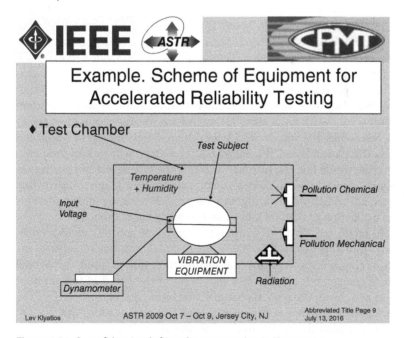

Figure 4.35 One of the visuals from the presentation in Figure 4.34.

machinery, and other areas of industry (complete machines as well as the components), thereby taking into account all the elements that could result in damage to components and protective films. This was particularly important, as many of these products were combined with mechanical wear or exposure to solar radiation. This is why the paper illustrates the modern essentials of accelerated corrosion testing as a component of ART/ADT.

Early in his career in the USA, his presentations and published written papers were aimed at different areas of industry. In fact, some of his early efforts began in electronic product testing, as demonstrated by the presentation for the IEEE Accelerated Stress Testing (AST) 1998 Workshop in Pasadena, CA (Figure 4.33). This presentation, "PRINCIPLES OF ACCELERATED RELIABILITY TESTING," was accepted only as a part of a panel presentation, because at that time the managers of this workshop were not familiar with his English publications in the electronics area. As Dr. Klyatis came from Moscow, he was unknown by Mr. Paul Parker, the chairman of this workshop, or the other workshop managers (Kirk A. Gray, President Accelerated Engineering, Inc., and others). A written paper from this event was published in the Proceedings IEEE Workshop on Accelerated Stress Testing, September 22–24, 1998, Pasadena, California.

A particularly memorable experience resulted from an IEEE workshop that included a visit of this meeting's attendees to NASA's Jet Propulsion Laboratory (JPL). The NASA staff at JPL demonstrated the new research station, which NASA planned to shortly sent to Mars. This moment is described in detail in the book published by John Wiley & Sons in 2012 [1]. Earlier, he had visited Kansas State University's research laboratory, which provided the testing of this Mars research station. During this visit he expressed concern that their testing was not accurately simulating the conditions on Mars. JPL staff responded that they were satisfied with their testing, and they were confident that the unit would work for at least 90 days, which was the required warranty period. A few weeks afterwards NASA sent the research station to Mars, and after 3 days the station failed. JPL called L.K. and asked why he had predicted that the station would fail. His response was that the testing was not accurately simulating the conditions on Mars; therefore, reliability and durability testing results were not truly predictive, and did not provide accurate information for successful prediction of the reliability and durability this station.

The main factor was that the temperature on Mars not only changes during the day from approximately −100 °C to +100 °C, but that sometimes the temperature changes very rapidly. The testing took into account the overall temperature change, but did not take into account the speed of temperature change and as a result, the station's batteries died prematurely.

Figures 4.34 and 4.35 show two visuals from his presentation "Specifics of accelerated reliability/durability testing" at the IEEE Workshop Renewable Reliability, ASTR 2009, Jersey City, NJ Hyatt Regency, USA, 09/16/2009.

Working as an SAE International representative to the Elmer A. Sperry Board of Award provides the opportunity to research outstanding contributions, including those in the area of reliability prediction implementation. A board member can work to sponsor a nomination for award for innovations that are outstanding scientific or technical achievements in the mobility field. The Elmer A. Sperry Board of Award consists of representatives from six mobility-related engineering societies, specifically the American Society Mechanical Engineers, the American Institute of Aeronautics and Astronautics, the Institute of Electrical and Electronic Engineers, the Society of Naval Architects and Marine Engineers, the American Society of Civil Engineers, and the Society of Automotive Engineers. This award, of which only one can be awarded per year, is given "… in recognition of an engineering contribution, which through application, proven in actual service, has advanced the art of transportation, whether by land, sea, air, and space." Dr. Klyatis made several presentations to the board, and he prepared the nomination of Dr. Zigmund Bluvband and Dr. Herbert Hecht for the Sperry Award. Their work in the development and implementation of novel methods and tools for the advancement of development and safety in transportation resulted in their becoming the 2011 recipients of this prestigious award. Figure 4.36 shows the 2011 award being presented at the SAE 2012 World Congress. It should also be noted that in managing his own company Dr. Zigmund Bluvband continues to implement the advanced ideas on quality and reliability.

Figure 4.36 Elmer Sperry Award ceremony during SAE 2012 World Congress (Detroit). From left: SAE International President, Elmer Sperry Board of Award Chairman Richard Miles, professor Princeton University, Dr. Lev Klyatis, this award sponsor, award recipients Dr. H. Hecht and Dr. Zigmund Bluvband.

He did not restrict his contributions to the Sperry Board to the subject of quality and reliability. He also successfully co-sponsored the nomination of an award to Thomas P. Stafford, Glunn S. Lunney, Aleksei A. Leonov, and Konstantin D. Bushyev as leaders of the Apollo–Soyuz mission and as representatives of the Apollo–Soyuz docking interface design team, and in recognition of the seminal work on spacecraft docking technology and the international docking interface methodology. This was in fact the first time the Sperry Award was presented for an achievement in aerospace.

Figures 4.37 and 4.38 respectively show the Elmer Sperry award recipients and sponsors for the Apollo–Soyuz project and Dr. Klyatis standing in front of the Apollo–Soyuz at the National Air and Space Museum in Washington, DC. The award to Dr. Bluvband and Dr. Hecht was impressive as most of the committee members relate better to award candidates in the area of product design, making awards for areas such as reliability and durability more difficult. A very important factor in advancing the implementation of reliability testing and prediction has been providing services as a consultant and seminar instructor for companies and organizations (especially large ones). For example, Box 4.5 shows the contents of a seminar for Ford Motor Company to

Figure 4.37 Group of recipients and sponsors of Elmer Sperry award for Apollo–Soyuz project during NASA Award Ceremony (Washington, DC). From left: Glynn Lunney, chair of Apollo project; Richard Miles, co-sponsor of award, professor Princeton University, Elmer Sperry Board of award member; General Thomas Stafford, chair of Apollo team; Lev Klyatis, co-sponsor of award, Elmer Sperry Board of award member.

Figure 4.38 Lev Klyatis in the National Air and Space Museum in Washington, DC, in front of the joined Apollo–Soyuz spacecraft (left is Apollo, right is Soyuz).

describe the new approach to reliability testing. One outgrowth of the positive reactions to this seminar and Wiley book [1] has been that, since 2012, SAE International has organized and provided a special technical session "Trends in Development of Accelerated Reliability and Durability Testing Technology" (IDM300), as part of each World Congress's sessions group "Integrated design and manufacturing."

Box 4.5 TRENDS IN DEVELOPMENT OF ACCELERATED RELIABILITY TESTING (Seminar for Ford Motor Company, April 2011)

Instructor Dr. Lev Klyatis (Director of Quality & Reliability ERS Corporation, Head of Reliability Department ECCOL, Inc.)

- Development techniques and equipment for more accurate simulation of real life input influences.
- Step-by-step development less expensive equipment for simultaneous combination of real life basic input influences.
- Development accelerated reliability testing, which offers the possibility to obtain directly the information for accurate prediction of reliability.
- Rapid obtaining accurate information for analysis the reasons of degradation mechanism and failures.
- Development the product quality through Accelerated Reliability Testing. It is specific.

(Continued)

Box 4.5 (Continued)

- Development of accelerated analysis of the climate influence on the new product reliability.

What is ART development? It is when you have more opportunities for:

(a) rapid finding of product elements that limit the product's quality and reliability;
(b) rapid finding of the reasons for the limitations;
(c) rapid elimination of these reasons;
(d) rapid elimination of product over-design (cost saving) to improve the product quality and reliability.
(e) increasing product quality and reliability, therefore resulting in a longer warranty period.

It is essential to simulate real life input influences on the product accurately for ART. If we cannot simulate real life influences accurately, we cannot perform ART and rapidly improve our product reliability.

The author is working in this direction and describes below how one can use it. One can use the above for development technology of ART

1. Determination the failures that limited the product reliability and quality.
2. Finding the location and dynamic of mechanism development of the above failures (degradation).
3. Finding the reasons of the above failures.
4. Elimination of these reasons.
5. Increasing of product's reliability and quality.
6. Increasing the warranty period of the product.

For the above ART implementation, one needs accurate simulation real life influences on the actual product.

A very important aspect of expanding the knowledge and implementation of successful reliability prediction has been international involvement by Russia, the USA, and world organizations in reliability and testing. Participation in this global experience helped L. K. to better understand the difference between Russian, American, and other nations' philosophies and practices relating to engineering and reliability. Unfortunately, it also revealed that many scientists, including many who had emigrated from the former Soviet Union, have difficulty adopting to the differing cultures in reliability and durability testing and prediction even though they worked for many years in America. Still, Dr. Klyatis firmly believes that working with, and discussions with, other world experts expands one's knowledge and open's up possibilities to improve and implement technology for reliability testing and successful prediction.

To date, the most effective method of disseminating these theories on the advancements in reliability testing and prediction has been through seminars, primarily with large industrial companies. And through these seminars and in visiting many industrial companies and research centers it was possible by studying the existing situations in reliability testing by many companies that Lev Klyatis was able to develop the way for his approach to be more easily implemented in practice. Examples include the invitations to seminars for Thermo King Corporation during the SAE World Congress in 2010, for Ford Motor Company during the 2011 World Congress (see Box 4.5), Nissan, and for Black & Decker Co.

The experience gained from these activities proved to be instrumental in helping to understand the direction that needed to be developed to advance the implementation of successful prediction of product reliability, and the need to take into account the specific nature of, and the cultural differences of different companies and organizations in Western, Eastern, and Asian countries.

Other activities include updating numerous presentations from 2010 through 2018, work as an ASQ Council of Reviewers member, reviewing numerous book proposals and books for different world publishers, including drafts of books, book reviews, paper reviews for SAE World Congresses and ASQ Congresses, and work for the New York State Assembly and the American Federal Government. All of these activities were involved in advancing the development and implementation of the successful prediction of product reliability.

The questions posed during these presentations also served to provide valuable feedback enabling better understanding of how to further the development of new directions for the successful prediction of product reliability. Figure 4.39 presents a typical meeting announcement.

Another example of how serving as chairman of the Technical Session for SAE World Congresses helped to implement his direction in successful prediction of product performance occurred when the managers from the Test Laboratory Ashok Leyland (India) submitted their paper's abstract on reliability testing for the 2017 SAE World Congress. The abstract, which was to be followed by the written paper, was submitted for the technical session "IDM300. Trends in Development of Accelerated Reliability and Durability Testing Technology." In his review of the abstract Lev Klyatis made recommendations that they change the title and contents, and described what the changes should be and why. The paper's authors had read and had a good understanding his book *Accelerated Reliability and Durability Testing Technology* [1]. The result of these discussions and recommendations was an updated written paper entitled "ACCELERATED COMBINED STRESS TESTING OF AUTOMOTIVE HEAD LAMP RELAYS", and included references to the book, and included recommended changes to the paper's content. And continuing the dialogue during the SAE 2017 World Congress (Figure 4.40), they had discussion with Dr. Klyatis on how to better use ART in their company.

SAE Metropolitan Section & NAFA New York-Intercounty Chapter

Double **Header**

Predicting Product Performance -- Safety, Quality, Reliability and Maintainability
&
Social Media Applications For The Fleet Professional

Joint Activity Meeting at the Port Authority Bus Terminal Times Square Room

We are excited to announce the first meeting of the New Year featuring two presentations. The morning presentation by Met Section's own Dr. Lev Klyatis will cover the topic of Successfully Predicting Product Performance with attention to safety, quality, reliability and maintainability. Lev has authored three books on quality, and his fourth book to be published by SAE is due out shortly. Lev is a world class expert on this subject.

The second presentation will be an introduction to social media for those of us who are not familiar with the benefits and practical applications of social media as a tool for the fleet professional. Two of our younger Governing Board members, Jesse O'Brien & Varuna Sembukuttige, will be doing this presentation.

Event Details:

Speakers:	*Predicting Product Performance* -- Dr. Lev Klyatis
	Social Media -- Jesse O'Brien & Varuna Sembukuttige
Date:	**Thursday February 4, 2016**
Location:	**Port Authority Bus Terminal – Times Square Room** *(Second Floor, South Wing, Next to Drago's Shoe Repair)*
Address:	**625 Eighth Avenue and 41ˢᵗ Street, Manhattan, NY**
Registration:	**10:00 am – 10:30 am**
Presentation *Predicting Product Performance:*	**10:30 – 12:00**
Lunch:	**12:00 am - 12:45 pm**
Presentation *Social Media*	**12:45 – 1:45**
Cost:	**$30.00 Members/Guests**
	SAE Students/Retirees - $15.00

JAMES REINISH
Chair
Port Authority of NY & NJ
Tel: (201) 216-2340
Email: jreinish@panynj.gov

ARTHUR KAPPEL
Vice Chair
Cablevision
Tel: (516) 803-2373
Email: akappel@cablevision.com

TOM LUBAS
Secretary
Honorary Member
Tel: (908) 568-3267
Email: lubas@comcast.net

DON RITTENHOUSE
Treasurer
Honorary Member
Tel: (516) 223-1213
Email: donritt@optonline.net

Figure 4.39 Meeting announcement for Dr. Klyatis presentation for engineers and managers of two societies in New York: SAE International Metropolitan section and NAFA's New York intercounty chapter.

Figure 4.40 SAE 2017 World Congress (WCX17), April 4, Detroit. Lev Klyatis (IDM300 technical session Chairman) hands Certificate In Recognition for Speaker Obuli Karthikeyan (Deputy Manager Component Test Lab, Ashok Leyland, India).

As published in [16], "Dr. Klyatis's 7 works in 39 publications in 1 language and 1,054 library holdings". This shows how wide L. K. publications in reliability prediction and testing are in usage.

4.4   Implementation of Reliability Prediction and Testing through Citations and Book Reviews of Lev Klyatis's Work Around the World

The following provides a partial list of publications throughout the world in which Dr. Klyatis work has been cited and references that may be of assistance in successful implementation of reliability testing and prediction. The following list gives examples of areas in which citation of his teams work has occurred:

- improving various product reliability issues in transportation refrigeration;
- marine energy converters;
- model aircraft and launcher controllers;
- statistical processing of accelerated life-cycle data;
- development of product models;
- durability testing of various products;
- improving quality testing and reducing customer complaints;
- improving wear and life characteristics of composite products;
- development of various aspects of reliability;
- development of various aspects of reliability testing;
- development of reliability assessment;
- development of electronic systems reliability;
- renewable and sustainable energy.

Citations

- Joseph Mannion. Project: Computer Integrated Testing for the Transport Refrigeration Systems Industry. Mayo Institute of Technology. DATE: 7/9/2000:

  > One important aspect of Accelerated Testing is the step-by-step strategy that can help obtain accurate and rapid initial information as results of product testing for reliability problem solving [6, 7] (Lev M. Klyatis, 1999).

- Philipp R. Thies, Lars Johanning, George H. Smith. Towards component reliability testing for Marine Energy Converters. EMPS – Engineering, Mathematics and Physical Science Renewable energy research group, University of Exeter Cornwall Campus, Treliever Rd, Penryn, TR10 9EZ, UK. Preprint submitted to Ocean Engineering May 26, 2010:

  > The type of test can be further distinguished, depending on how accurate the field loads are replicated and to what extent they are accelerated (Klyatis and Klyatis, 2006):
  > o Field testing of the actual system under accelerated operating conditions
  > o Laboratory testing of actual system through physical simulation of field loads
  > o Virtual (computer-aided) simulation of system and field loads.

- Allen Zielnik. Atlas Materials Testing Technology LLC. Validating photovoltaic module durability tests. Solar America Board for Codes and Standards. Ametek Corporation. www.solarabcs.org. July 2013:

  > Klyatis (Klyatis, 2012) further points out that the vast majority of literature references to reliability tests are actually true durability tests, or stop short of failure at a finite duration as a qualification test. Despite extensive use of the term "reliability," most of what is being done in the PV industry at best falls under the realm of trying to assess "durability.
  >
  > As Klyatis notes, many types of environmental influences act on a product in real life. Some influences are studied, but many are not. The factors responsible for the influences of environmental stresses in the field are very complicated. One of the most complicated problems is the integrated cause-and-effect relationship of different factor steps including stress influences, effect on output parameters, and degradation. In accelerated testing, we have two primary means of test acceleration. The first is overstress where one or more levels of a stress condition (such as temperature cycling) are applied at levels in excess of the intended normal service use. The test results are then used to

extrapolate estimated performance at the normal stress level. Care must be taken not to exceed the stress strength of the product and induce unrealistic failure.

- Lisa Assbring, Elma Halilović. Improving IKEA's Quality tests and management of customer complaints—kitchen fronts and worktops. Master of Science Thesis MMK. 2012:25 MCE 275 KTH Industrial Engineering and Management Machine Design SE-100 44 Stockholm, Sweden:

  With AT, results that would require years of field testing may be obtained in days or weeks (Klyatis & Klyatis, 2005).

  There are three general methods of accelerated testing. The first method is to test a product under normal field conditions and subject it to the same impact that would happen in real life (Klyatis & Klyatis, 2005). When the hinge on a kitchen front is tested, it is subjected to the same impact of being opened and closed that would happen during normal use in a kitchen.

  The second method is to use special equipment in a laboratory to simulate the influences on a product (Klyatis & Klyatis, 2005). An example of this is when the wear and tear of years of usage of a worktop is simulated using a tool that causes abrasion on the surface. Since the influence on the product is simulated with a tool, the conditions do not correspond to normal field conditions. Accurate test results require accurate simulation of the field input.

  The third method is to use computer software to simulate the effect on the product. The accuracy of this method depends on how accurately both product and the influences on the product can be simulated (Klyatis & Klyatis, 2005).

  The approach of AT is to either accelerate the use-rate or the product stress. Use-rate acceleration can be used when a product is normally not continuously in use, and it is assumed that its lifetime can be modeled in cycles. These cycles are accelerated during testing and the product's lifetime estimated. This is sometimes referred to as ALT (Klyatis & Klyatis, 2005). For example, the expected number of times the kitchen front will be opened during its lifetime can be calculated, and by accelerating these cycles the effect they will have on the hinge can be tested.

  In accelerated stress testing the stress factors that cause product degeneration are accelerated. For example, temperature or concentration of chemicals may be increased in comparison to normal usage of the product. In the wear test of the worktop both the stress caused by the abrasion and the cycles are accelerated. Stress acceleration is often used to identify stress limits and design weaknesses. Combined stresses, such as temperature and vibration, are often especially efficient in accelerating failure of a product (Klyatis & Klyatis, 2005).

The equipment used during testing will also affect the results, as it can often only simulate one type of input while the product during normal conditions may be exposed to several simultaneously (Klyatis & Klyatis, 2005).

- Gino Rinaldi. NSERC Research Fellow. A Literature Review of Corrosion Sensing Methods. Technical memorandum. Defense Research and Development Canada. September 2009. cradpdf.drdc-rddc.gc.ca/PDFS/unc102/ p533953_A1b.pdf

  A. 13.40 Klyatis (2002): Establishment of Accelerated Corrosion Testing Conditions
  The author of this work developed techniques for accelerated corrosion testing (ACT) and implemented them. The main goal of ACT is rapid product improvement, lower warranty costs, lower life-cycle cost, and improved reliability through accelerated corrosion testing. [168]

- David Booth. The man who kills Kias for a living. PressReader. The Province: 2016-02-01. Canada. www.pressreader.com/canada/the province/

  ... It's part of the goal – conducted on proving grounds from Finland to Dubai (Kia, in fact, has two more in South Korea) – of producing an endurance test that Lev Klyatis, author of Accelerated Reliability and Durability Testing Technology, describes as being 150 times more intense than normal driving.

- Accelerated stability testing: Topics by Science.gov. *Sample records for accelerated stability testing*. Conditions of environmental accelerated testing (Citation 2). Microsoft Academic Search:

  Strategic and tactical bases are developed for conditions for environmental accelerated testing of a product using physical simulation of life processes. These conditions permit rapid attainment of accurate information for reliability evaluation and prediction, technological development, cost effectiveness and competitive marketing of the product, etc. (L. M. Klyatis).

- Bulletin for the International Association for Computational Mechanics (IACM). #20, January 2007. Expressions 20.4.qxd:Expressions 20.4.qxd – Cimne. www.cimne.com/iacm/News/Expressions20.pdf

  Book Report
  Accelerated Quality and Reliability Solutions
  Lev Klyatis & Eugene Klyatis (Eds.)
  Edited by ELSEVIER, EUR 136

Drawing of real world issues and with supporting data from industry, this book overviews the technique and equipment available to engineers and scientists to identify the solutions of the physical essence of engineering problems in simulation, accelerated testing, prediction, quality improvement and risk during the design, manufacturing, and maintenance stages. For this goal the book integrates Quality Improvement and Accelerated Reliability/Durability/Maintainability/Test Engineering concepts.

The book includes new unpublished aspects in quality: – complex analysis of factors that influence product quality, and other quality development and improvement problems during design and manufacturing; in simulation: - the strategy for development of accurate physical simulation of field input influences on the actual product – a system of control for physical simulation of the random input influences – a methodology for selecting a representative input region for accurate simulation of the field conditions; in testing: – useful accelerated reliability testing (UART) – accelerated multiple environmental testing technology – trends in development of UART technology; in studying climate and reliability; in prediction – accurate prediction (AP) of reliability, durability, and maintainability – criteria of AP – development of techniques, etc.

- Youji Hiraoka and Katsumari Yamamoto, Jatko Ltd (Japan); Tamotsu Murakami, University of Tokyo; Yoshhiyuki Furukawa and Hiroyoki Sawada, National Institute of Advanced Industrial Science and Tech. (Japan). Method of Computer-Aided Fault-Tree Analysis (FTA) Reliability Design and Development: Analyzing FTA Using the Support System in Actual Design Process. SAE 2014 World Congress. SAE Technical paper 2014-01-0747:

In reliability engineering the ART/ADT strategy components is advocated by Klyatis Lev [21]. This is a method of product quality guaranteed by testing process that uses FMEA and FTA to investigate the testing conditions of ART/ADT.

Published Reviews of Dr. Klyatis's Publications in English

There are many published reviews of Lev Klyatis's (12) published books, and over 250 articles, and papers. Four of the books have been in English. One can see below several from these published reviews

1. Book review, published in *The Journal of RMS (Reliability, Maintainability, and Supportability) in Systems Engineering*, DoD, Winter 2007/2008; see Figure 4.41.

BOOK REVIEW

<u>Accelerated Quality and Reliability Solutions</u>
by Lev M. Klyatis, PhD and Eugene L. Klyatis, MS
Published by Elsevier, Oxford, UK, 2006

by John W. Langford, BS, MS, CPL, CCM

Overview

This book describes accelerated quality and reliability solutions of the physical essence of engineering problems during the design, manufacturing, and maintenance stages of systems and equipment. The premise of the book is based on integration of Quality Improvement as well as Accelerated Reliability, Durability, Maintainability and Test Engineering concepts. The book introduces a complex technique which consists of five basic disciplinary groupings:

1) physical simulation of field input influences;
2) useful accelerated reliability testing;
3) accurate prediction of reliability, durability, and maintainability;
4) accelerated quality development and improvement in manufacturing and design; and
5) safety aspects of risk assessment.

Figure 4.41 Book review, published in *The Journal of RMS (Reliability, Maintainability, and Supportability) in Systems Engineering*, DoD, Winter 2007/2008.

2. Published review in journal *Total Quality Management and Business Excellence*, Taylor & Fransis Group, Volume 17, Number 7, September 2006, UK; see Figure 4.25.
3. Dr. Russ Vacante. RMS Partnership President. *The Journal of Reliability, Maintainability, & Supportability in Systems Engineering* (Introduction). DoD, Summer 2012.

> Book Review. *Accelerated Reliability and Durability Testing Technology*. WILEY. 2012.
> Accelerated Reliability and Durability Testing Technology, authored by Lev M. Klyatis and published by Wiley, is a interesting

The strength and credibility of the book is reinforced by the accuracy reflected in true simulations of field situations along with effective application of mathematical methodologies and models derived primarily from statistical, calculus, and algebraic algorithms.

Textual Contents and Topical Organization

Chapter 1: Accurate Physical Simulation of Field Input Influences on the Actual Product: This chapter describes the strategy, basic concepts, criteria, and methodological aspects for the development of accurate physical simulation of the governing field input influences as well as the required system of control of the simulation process. For this purpose, the techniques of substitution of artificial media for natural technological media and a description of how the climate influences reliability are included.

Chapter 2: Useful Accelerated Reliability Testing Performance: This chapter provides a general analysis of different approaches to accelerated testing methods as well as the specifics of useful accelerated reliability testing (UART) methodology. It also describes the basic UART on the basis of physical simulation as discussed in Chapter 1, which makes it possible to derive information for accurate quality, reliability, durability and maintainability (RDM) predictions as well as a decrease in safety risks.

Figure 4.41 (*Continued*)

and eye opening excursion on how to conduct, properly, accelerated reliability and durability testing (ART/ADT). New concepts an ideas are centered on close simulation of the conditions that equipment are exposed to out "in the field". The author maintains that simulation of the actual product environment will improve product reliability, maintainability, and safety, reduce time of product design, manufacturing, and usage. Thus, accurate simulation of the environment a system is exposed to during actual use, in conjunction with the use of the proper test equipment, is the key to successfully conducting ART/ADT.

Duplicating of the product actual environment may seem intuitively easy to accomplish at first, however, it is seldom, if ever, practiced within many test communities, organizational cultural paradigms with respect to what constitutes reliability testing are often deeply entrenched and resistant to change. Additionally, Lev Klyatis's book has a wide range of applicability. His holistic approach to accelerated reliability and durability testing will resonate with testing professionals as well as aspiring test engineers. His highlights of the drawbacks associates with current testing practices offer a sound basis for the improved testing alternatives he proffers. The reader walks away with cost effective arguments that can be offered to decision makers to explain how changes to testing procedures and training, along with the purchase and use of the proper test equipment, can reduce cost and improve system of reliability and safety.

The format, information and exploration contained in this book are well crafted, easy to understand, and illustrative. For example, the opening chapters of the book challenges the current contents of reliability testing institutionalized in many government and non-government organizations. The author provides inside to a conversing argument for the idea that current testing practices do not significantly improve product and system reliability. In fact, the organizations that fail to adopt the author's recommended early on life cycle and holistic protocols to reliability testing may increase their overhead and product/system cost.

Members of the testing community who acquire the author's perspective on reliability and durability testing will understand what they need to do to improve system reliability. The simulation of "real world" environmental conditions, while seemingly at first more challenging will, in the end, prove to be extremely rewarding in terms of improving reliability and safety while lowering cost(s). Equally important is the need for the test community to grasp a fundamental concept: those performing the testing must be well trained and experienced.

The closing chapters of this book focus on ART/ADT new technology as a credible source of information for accurately predicting product quality, reliability, safety, maintainability, durability, and life cycle cost. This information is followed by a discussion on critical need their standardization, which professional societies can be expected to encourage. Additionally, the author introduces a strategy for prediction for improving predication accuracy during development and provides models and solutions, all of which have their foundation based on the need to simulate the actual use environment. The author reinforces his holistic on human factor and safety consideration that current testing publications often fail to fully address

Mr. Klyatis's book bring a refreshing, easy read style to the subject. He does a suburb job of demystifying the relationship between ART/ADT and cost by introducing an approach that moves reliability testing from an art to a science. The reader(s) of this book will learn of

Conclusion

This book is a must for engineering professionals – especially those from the logistics engineering and systems engineering communities – to have in their reference libraries. For the convenience and cost benefit of the reader, the authors have amalgamated the essential material from the critical sectors of reliability, maintainability, quality assurance and supportability into a single compendium of valuable guidance to fulfill the stated purpose of the book. The authors are to be highly commended for their clear and innovative approach. ★

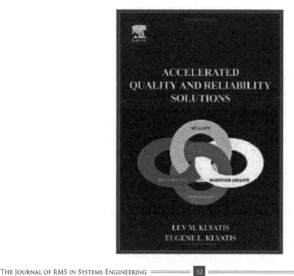

Figure 4.41 *(Continued)*

a new approach to accurate simulation, testing, and prediction that will benefit them and their organization(s).

4. Reliability Review. ASQ. *The R & M Engineering Journal*, **4.** Volume 26, Number 4, December 2006, Page 29.

RR BOOKSHELF
Accelerated Quality and Reliability Solutions
by Lev M. Klyatis and Eugene L. Klyatis. Elsevier, Oxford, UK, 2006

Most professionals entering the workforce today are required to analyze situations, identify problems, and provide solutions for improved performance. This book is particularly salient for engineers and managers for practical improvements in simulation, accelerated reliability testing, and reliability/maintainability/durability prediction. The content of this book is structured with five chapters

1. Accurate physical simulation of field input influences on actual product.
2. Useful accelerated reliability testing performance
3. Accurate prediction of reliability, durability and maintainability from reliability testing results.
4. Practical quality development and improvement in manufacturing and design
5. Assessment of safety risk.

This updated and expanded edition presents many different solutions in an easyto-follow structure: a description of the tool, its purpose and typical applications, the correct procedure is illustrated, and a checklist to help one apply it properly, The examples, forms and templates used are general enough to apply to any industry or market.

The layout of the book has been designed to help speed your learning. The book is suited for employees working in Quality and Reliability fields in any type of industry. It includes a variety of data collection and analysis tools, tools for planning, and process analysis.

Important benefits of this book are:

- New concepts which can help to solve earlier inaccessible problems during design, manufacturing, and usage;
- Can help dramatically reduce recalls.
- Develops a new approach to improving the engineering culture for solving reliability and quality problems;
- Prescribes a practice for improving reliability/maintainability/durability.

This more than 500 page book is an excellent addition to the library of every industrial, service, processing, high tech, consultant, training company, as well as every engineer and manager.

Reviewed by:
Dak K. Murthy, P.E., C.Q.E.
Manager, Quality Assurance, NJ Transit and Chair of ASQ NY/NJ Metropolitan section.

4.5   Why Successful Product Prediction Reliability has not been Widely Embraced by Industry

There are several basic aspects as to why successful prediction and testing of product reliability has not been widely embraced by industry. These include:

- As demonstrated in Chapter 1, most current solutions of reliability prediction relate to theoretical aspects with artificial examples.
- These solutions are not closely connected with obtaining accurate initial information from the source for the prediction calculation for the product.
- The usual source for obtaining reliability information is testing results, using traditional testing methods and equipment that do not accurately simulate the real-world conditions. For example, proving grounds testing, which has been in use for about 100 years, has developed very slowly; and does not accurately simulate many real-world input influences (multi-environmental, electronic), as well as other important factors such as human factors and safety problems. A similar situation exists with laboratory testing. As a result, proving grounds and laboratory testing results are often very different from real-world operational results.
- There is a lack of accurate and consistent definitions of terms used in testing. Therefore, the use of misleading or incorrect terminology is often used, resulting in incorrect assumptions or conclusions. For example, vibration testing or proving grounds testing is often incorrectly called "reliability testing," even by professionals in this field.
- It is difficult to find literature that presents correct and accurate definitions and terminology in the field of reliability and durability testing, ART, and ADT that accurately simulate real-world conditions.
- Similar problems exist in finding accurate literature on how one can simulate field conditions accurately.
- Often there is a reluctance by management to invest monies in advanced reliability prediction solutions in industry. This is especially true when these improvements require investment in new technology or testing equipment.
- Professionals in design, research, and testing often prefer simulation of separate (or using only partial field influences), because it appears to be simple and inexpensive. But too often they are only counting the direct cost of research and testing, without taking into account subsequent costs resulting from the failure to provide accurate simulation of the product. As a result of their "savings" by simple and inexpensive testing, there are recalls, complaints, and less profit.

- Crash testing is also an area where results are not always predictive. In the real world, crashes depend on many input variables, including human factors and other safety problems which usually are not taken into account in the structured crash test.
- For these reason, and others, successful prediction of product's reliability has not been widely embraced by industry.

Further details concerning these problems and their solutions can be found in this book, as well as in author's over 250 publications; including Refs [8–15].

References

1 Klyatis L. (2012). *Accelerated Reliability and Durability Testing Technology.* John Wiley & Sons, Inc., Hoboken, NJ.
2 Klyatis L. Development standardization "glossary" and "strategy" for reliability testing as a component of trends in development of ART/ADT. In *SAE 2013 World Congress*, Detroit, paper 2013-01-0152.
3 Klyatis L. Development of accelerated reliability/durability testing standardization as a components of trends in development accelerated reliability testing (ART/ADT). In *SAE 2013 World Congress*, Detroit, Paper 2013-01-151.
4 Klyatis L. (2016). *Successful Prediction of Product Performance. Quality, Reliability, Durability, Safety, Maintainability, Life Cycle Cost, Profit, and Other Components.* SAE International, Warrendale, PA.
5 Klyatis L, Klyatis E. (2006). *Accelerated Quality and Reliability Solutions.* Elsevier.
6 Anon. (1990). Editorial article. Interview with Klyatis L. M., Sc.D. High quality rigging for testing service. *Tractors and Farm Machinery* (November). USSR Department of Automotive and Agricultural Industries.
7 *SAE* 2013 *World Congress. Achieving Efficiency. Event Guide.* SAE International.
8 Klyatis L. (2006). Introduction to integrated quality and reliability solutions for industrial companies. In ASQ World Conference on Quality and Improvement Proceedings, *May* 1–3, Milwaukee, WI.
9 Klyatis LM. (2002). Establishment of accelerated corrosion testing conditions. In *Reliability and Maintainability Symposium (RAMS) Proceedings*, Seattle, WA, January 28–31; pp. 636–641.
10 Klyatis LM, Klyatis E. (2001). Vibration test trends and shortcomings, part 1. *The R & M Engineering Journal (ASQ): Reliability Review* **21**(3).
11 Klyatis L. (1985). *Accelerated Testing of Farm Machinery.* Agropromisdat, Moscow.

12 Klyatis L. (2017). Why separate simulation of input influences for accelerated reliability and durability testing is not effective? In *SAE 2017 World Congress*, Detroit, April, paper 2017-01-0276.

13 Klyatis L. (2016). Successful prediction of product quality, durability, maintainability, supportability, safety, life cycle cost, recalls and other performance components. *The Journal of Reliability, Maintainability, and Supportability in Systems Engineering* (Spring): 14–26.

14 Klyatis L. (2009). Accelerated reliability testing as a key factor for accelerated development of product/process reliability. In *IEEE Workshop Accelerated Stress Testing & Reliability (ASTR 2009)*, October 7–9, Jersey City [CD].

15 Klyatis L, Walls L. (2004). A methodology for selecting representative input regions for accelerated testing. *Quality Engineering* **16**(3): 369–375.

16 Klyatis, Lev M. [WorldCat identities], worldcat.org/identities/locn-1102003091931

Exercises

4.1 Show the basic financial factors that need to be included in the cost/benefit analysis for implementation of the new approach to ART and reliability prediction.

4.2 Describe examples of reliability testing implementation.

4.3 Describe examples of reliability testing and prediction implementation.

4.4 Describe how ART was implemented in ASAE TC456 standard Test and Reliability Guidelines.

4.5 Describe how the new approach to reliability testing was implemented in SAE Aerospace standards JA1009 Reliability Testing.

4.6 Provide the definitions for the following terms from the standard JA1009 Reliability Testing – Glossary:
   o accelerated testing;
   o accelerated reliability testing (ART) or accelerated durability testing (ADT);
   o accurate prediction;
   o accurate system of reliability prediction;
   o accurate physical simulation.

4.7 Describe the basic concepts of SAE standard JA1009 Reliability Testing – Strategy.

4.8 What is the basic content of Richard Rudy's (DaimlerChrysler) book review, published in the *Journal of Total Quality Management and Business Excellence* (UK)?

4.9 Provide some reasons why ART or ADT is actually more economical than single testing with simulation of separate influences?

4.10 Describe how Kamaz, Inc. began to implement reliability testing and reliability prediction?

4.11 Why is it incorrect to describe "vibration testing" as "reliability testing or durability testing" of the product?

4.12 Why did Dr. Klyatis predict that NASA's Mars research station would fail?

5

Reliability and Maintainability Issues with Low-Volume, Custom, and Special-Purpose Vehicles and Equipment
Edward L. Anderson

5.1 Introduction

There remain today many industries where reliability testing is not developed or used in the products they produce and deliver. In these industries, testing, if performed at all, consists of only functional, performance, or operational testing at the factory, followed by some degree of ad hoc product evolution through real-world operational experience by the customer. This is especially true when dealing with manufacturers of low-volume, unique, custom or special-purpose vehicles and equipment. Testing that may be done is mostly functional performance testing. This author's experiences are primarily with custom automotive vehicles typically production-line light-duty vehicles modified for special vocational applications, medium- and heavy-duty trucks, construction equipment, emergency response vehicles for fire, police, and EMT services, standby or emergency generators and fire pumps, and work or patrol boats. Figures 5.1 through 5.6 depict typical equipment of this type. In his career with the Port Authority this author has been the responsible engineer for the acquisition of over 5,000 such vehicles with procurements of vehicles and equipment costing over a half billion dollars.

Most of these units are of limited production and frequently have minimal or no true reliability testing. But these units do have some common elements. They are often mission-critical assets necessitating a high degree of reliability and maintainability. They are complex units and require operator and maintainer proficiencies. They are expected to work over a long service life. Finally, they generally are expensive and difficult to replace.

Another factor is the risks associated with acquiring such units from a vendor that has poor reliability or maintainability can be very high—especially when there are serious consequences for not having unit(s) in service, there

Reliability Prediction and Testing Textbook, First Edition. Lev M. Klyatis and Edward L. Anderson.
© 2018 John Wiley & Sons, Inc. Published 2018 by John Wiley & Sons, Inc.

Figure 5.1 Typical airport snow and ice control equipment.

Figure 5.2 Emergency generator rooftop installation.

are legal liabilities for not having operational units, there are negative public relations implications, or similar consequences to the equipment owner or service provider. Consider the consequences to your organization should an ambulance, police, or fire response vehicle break down and fail to respond to a life-safety incident. Unfortunately, too often, organizational procurement practices may not allow selection of any offering other than lowest price.

Figure 5.3 Aircraft fueling cart (tow type).

Figure 5.4 Typical airport runway deicer vehicle.

Figure 5.5 Elevating platform truck.

5.2 Characteristics of Low-Volume, Custom, and Special-Purpose Vehicles and Equipment

While there can be no question that the testing detailed in other chapters in this book provides the surest solution to providing customers with the reliability and maintainability desired in a product, and provides the best profitability to the manufacturer, it is also true that many products, especially in the truck and equipment fields, are of low production volume and have little or no field testing by the manufacturer. Where testing is done, it probably meets the minimum requirements of consensus standards such as stability

Figure 5.6 Emergency fire pump.

testing for man lifts and Aircraft Rescue and Fire Fighting (ARFF) vehicles [1] or pumping performance tests for fire pumpers. These tests typically are performance standards for a single set of test conditions and do not include widely different temperature or environmental conditions, so they are not truly real-world conditions as described in this book. (The user must also consider whether industry consensus standards are adequate or appropriate for their operational needs. If they are not, more stringent requirements may be contractually required.) Often, these units may be unique, and often the "manufacturer" is an assembler of varied commercially available components with little control over the design of these components. Rarely are these units produced on a true assembly line, but rather are assembled by skilled workers on a unit-by-unit basis. Usually, such equipment is also of significant dollar value, having a long expected service life. Often, they also provide a critical function supporting the end user's mission or business function.

Assuring reliability and maintainability in such situations is especially challenging, particularly to the customer who has few mechanisms to assure purchase of a quality product (often depending solely on the manufacturer's sales or marketing claims), and the results of receiving a poor product can be expensive or potentially disastrous: How will your organization be affected by an ambulance, or police emergency unit that breaks down during response, or a backup generator or fire pump that fails to start during a power failure? The costs, including monetary loss, image, and potential legal liabilities could be devastating. But the results to your business or organization can be as problematic with failures in more routine products, such as delivery trucks, plumbers, electricians, or other trade' mobile equipment.

Let us consider some of the attributes of such equipment, especially as related to the mobile equipment industry.

- *Long life/high cost/long lead time*   Equipment of this nature will typically have a long life (often greater than 10 years in service) and in many cases will become technologically obsolete rather than mechanically cost prohibitive as end of useful life approaches. Consider how technology has changed in the last 20 years, and it is easy to see that units that were state of the art then are lacking in many features we take for granted today: GPS, cell phones, backup, and dashboard cameras to name but a few. Emissions changes and fuel choices are very different today.

  In many situations, the acquisition process is complex, with the development of written specifications defining the intended functionality of the equipment, and detailing many specifics that are related to the operating needs, operating environment, regulatory concerns, environmental impacts, and other considerations. This, too, must be included in the acquisition lead time, and often entails a committee or work group to formalize these into needs and wants. The development of these requirements can easily add one to two more years to the acquisition process, bringing the total time from recognition of the need to acquire or replace a unit to 3–5 years.

  Unlike going to a store and buying an "off-the-shelf" product, or going to a car dealership and buying an automobile from inventory, acquisition of this equipment typically takes much longer, and the unit is produced to order. Typical lead times from placement of order to delivery of finished product can be over 2 years.

- *Strategies for assuring reliability and maintainability of low-volume, custom and special-purpose vehicles and equipment*   Although these units rarely have extensive or real-world testing, there are factors that can be employed to minimize the risks of acquiring poorly performing units, and to reduce the risks of operating substandard equipment. By using the tactics and strategies described in the following it is possible to minimize the risk of acquiring substandard units.

5.2.1   Product Research

Before beginning the procurement of new units, extensive research needs to be performed, preferably by a professional with engineering or technical expertise, to gain knowledge of the experience with the present generation (or a needs analysis for units that have no prior experience), changes that have occurred in the industry since the last procurement, and most importantly a look into the future to see if the operational needs are expected to change over the life of the new equipment. Research should involve operational management to define operational needs, machine operators, maintenance providers, and new equipment vendors.

5.2.2 Vendor Strength

A common feature on much of this equipment is a small vendor community (typically less than 10 manufacturers, and frequently less than five manufacturers). A key concern is whether the vendor has a robust engineering commitment. Frequently, sales staff will have little technical abilities, or answers to hard or difficult queries as to systems design or operational concerns. Vendors with a strong corporate engineering culture are more likely to be committed problem-solvers than organizations with weak engineering. (It is this author's observation that successful organizations become successful by offering new or unique products and are entrepreneurial by nature. But, unfortunately, many later morph into short-term profit- driven institutions and engineering and innovation are avoided or discouraged.) Unfortunately, this is a factor that generally is difficult to quantify, and organizational procurement policies may not allow consideration of the engineering strength of the manufacturer even though this factor is among the most critical in acquiring specialized or custom equipment.

5.2.3 Select a Mature Product

As a product is used in real-world service, problems become evident and solutions are developed. Even with the best efforts in engineering a new product or updated model, problems will be discovered without real-world testing. It is almost a certainty that purchasing the 100th unit will result in better reliability than the first production unit. The reality is that the customer is the real-world test track for this equipment. Quality vendors will learn from customer experiences and improve their products over time. Over time, it has been this author's experience (for vehicles as varied as police cars, snow blowers, airport rescue and fire fighting vehicles, and even new or updated model medium- and heavy-duty truck chassis) that the problems and failures occurring in the first year or two of a newly introduced model or product will be significant. And as these are new and unanticipated problems, solutions are rarely quick or simple, and often result in retrofit or modification to all units in the order.

5.2.4 Develop a Strong Purchase Contract

Because of the high cost of virtually all this type equipment, a written agreement, contract, or purchase order is required. This written document can provide the customer with clearly defined rights that can help assure product reliability and maintainability, and clearly define what level of performance is expected from the manufacturer or supplier. Performance requirements, defined "mean time to failure" requirements, operational limitations or restrictions, redress for nonperformance, and numerous other details need to be clearly specified and agreed to by both parties in this written document. There should be little ambiguity and unaddressed items in this written document, for the protection of both parties. Assuring product reliability and maintainability starts with a clear understanding of what is the expectation by the user.

5.2.5   Establish a Symbiotic Relationship

As previously indicated, this equipment generally does not undergo real-world testing. By the customer sharing their real-world experiences with manufacturers who are interested in, and committed to, improving their product the result can be a mutually beneficial relationship. However, for this to work there is a need for good communications between the parties, and trust that the information sharing will be used for improving the product and not for attributing blame or fault. It may be prudent to codify this agreement to share information with limiting associated liabilities.

5.2.6   Utilize Consensus Standards

Much of this equipment does fall under industry-accepted consensus standards, which, among other things, set minimum performance standards that manufacturers should meet or exceed. These should be the minimum standards acceptable, but this does not mean that the customer cannot set more stringent requirements. For example, ANSI Standard A92.2 for Vehicle-Mounted Elevating and Rotating Aerial Devices [2] permits the lifting of a tire or outrigger under the definition of stability. But, in this author's experience with several agencies, stability has been defined as prohibiting any outrigger or tire lifting from the ground. Of course, when the customer requires different requirements than that called for in the consensus standard, the costs associated with new performance testing will almost certainly be an incremental cost to the unit. (It should also be noted that earlier in this author's career it was common for the engineer to take exception to industry consensus standards when in his professional judgement there were valid reasons for the exception; however, due primarily to the legal liabilities in the USA, deviating from consensus standards today will rarely be supported and is quite risky.) While consensus standards often include performance measures and/or functional testing, it should be noted that most of these requirements are to assure uniformity in performance measures. It establishes uniform functional testing methods and results. They are generally not designed to duplicate all real-world operational conditions as needed for successful reliability

Figure 5.7 Author overseeing SAE ARP 5539 snow blower performance testing [3].

testing. Figure 5.7 shows this author during "SAE ARP5539 Rotary Plow with Carrier Vehicle" performance testing. This functional testing provides a uniform method of measuring the ratings for speed and tons per hour of snow removal by high-speed airport and highway snow blowers. It is also strongly recommended that organizations become participants in consensus standards committees. These groups generally have many manufacturers working on their standards, but need user volunteers to balance their committees and to be sure the developed standard meets the user's needs. Figure 5.8 illustrates consensus standards functional testing.

5.2.7 User Groups/Professional Societies

User groups and professional societies are valuable resources for sharing information between people and organizations with similar interests, problems and solutions. The American Public Works Association (Kansas City, MO), ARFF Working Group (Grapevine, TX), National Association of Fleet Administrators (Quincy, MA), the National Fire Protection Association (Quincy, MA), and SAE International (Warrendale, PA) are a few of such organizations.

5.2.8 Prerequisites

Prerequisites are a tool that can be used to preclude vendors with little or no experience in producing the requested units. It provides assurance that the

Figure 5.8 National Fire Protection Association 414 ARFF tilt table testing [1].

manufacturer (or in some cases the distributer) is experienced and knowledgeable on the product. Unfortunately, it can also exclude some vendors who may be offering a new or innovative product. Typical prerequisites would be a requirement to have manufactured a minimum number of similar or identical units in a specified time frame.

5.2.9   Extended Warranties

Extended warrantees are another tool to enhance the reliability of a unit. These can be applied to the entire unit, or to certain components. In medium- and heavy-duty trucks, it is common to have separate warrantees for the engine and the transmission. Lifetime warrantees against chassis, body, or frame rail cracking, corrosion, or permanent deformation can also be used to provide assurances on unit reliability. Generally, extended warrantees must be agreed to by the manufacturer, and they can be costly.

5.2.10   Defect/Failure Definitions/Remedies

Contractually defining defects or failures, and their remedies, can be used to define the expectations of the customer for the reliability of the unit. These are common in some industries, but not in all. A typical clause used for transit bus procurements is as follows.

Defects/Failures

The following shall be the design objectives for maximum frequency of in-service failures of the types defined, provided that preventive maintenance procedures specified by the Vendor are followed within the limits of practicability dictated by transit maintenance practice.

A. Class I: Physical Safety—A failure which leads directly to passenger or operator injury or represents a severe potential crash situation, an example of which is the loss of vehicle brakes. Mean distance between failures shall be greater than 1,000,000 miles, or the actual life of the bus.

B. Class II: Road Call—A failure which results in the interruption of service. An example of this is a vehicle breaking down during service. Mean distance between failures shall be greater than 20,000 miles.

C. Class III: Bus Change—A failure which requires removal of the bus from service during its assignment but does not cause an interruption in revenue service. An example of this would be full loss of HVAC or interior lighting systems or engine fault reducing engine power to a limp-in mode. Mean distance between failures shall be greater than 16,000 miles.

D. Class IV: Bad Order—A failure which does not require removal of the bus from service during the assignment, but degrades the operation or use of the bus, causing the failure to be reported by the operator. An example of this would be an inoperative lamp, public address system, or partial failure of the security recording system. Mean distance between failures shall be greater than 10,000 miles.

It should be noted that defining defects and failures requires knowledge of reasonable failure intervals to be enforceable.

5.2.11   Pre-Award and/or Preproduction Meetings

A key part of assuring the units will provide the required reliability and maintainability objectives can be the requirement to have pre-award and/or pre-production meetings. These meetings can be instrumental in assuring that all parties understand the needs and mission of the completed vehicle and the manufacturer's processes and deliverables. These meetings should involve all interested parties. While similar in nature, the pre-award meeting generally does not include fully developed technical details on the product, but is more of an assurance that the manufacturer and client have the same understanding of the proposed contract and deliverables. The pre-award meeting should not provide for negotiations or contract changes, but an assurance that what is being asked for will be provided. The preconstruction meeting is a more detailed meeting with full detailing of the technical aspects of the offering, and the proposed timeline for the deliverables. Minutes of both pre-award and preproduction meetings should be published, disseminated to all involved parties, and become part of the project records.

5.2.12 Variation

As presented earlier, these units are rarely produced on a true assembly line. As a result, variations or inconsistencies in the assembly of the unit are common. Different workers may do things slightly differently, minor differences may occur in the delivery of parts to the workstation, or human errors occur, such as, "I thought I tightened those bolts just before quitting time yesterday"; or "it's a lot easier to route this wiring harness different than the drawing shows." While most of these variations are benign and may go unknown through the product's life cycle, occasionally they can have serious results.

An actual example of this occurred when the wiring harness in the engine compartment of an order of a new transit bus was routed and supported with very slight differences—so slight it was difficult to see even after learning of the issue. Unfortunately, in the correct method the harness was supported to hold the harness an inch or two above the diesel engine's high-pressure injector fuel line. In some buses, a clamp to hold the harness was in a slightly different position and the full weight of the harness rested on the injector line. Over time the protective cover on the wiring harness wore through the injector line, developing a hole which sprayed diesel fuel throughout the engine compartment. The atomized fuel found an ignition source and resulted in the rapid spread of a fire that destroyed the bus. Fortunately, the bus had no passengers onboard at the time of the incident. The bus was equipped with an automatic onboard fire suppression system which discharged, and the driver and nearby personnel attempted to suppress the fire using handheld extinguishers, but the bus was a total loss (Figure 5.9). It took the efforts of several expert fire investigators to

Figure 5.9 Bus lost due to fire.

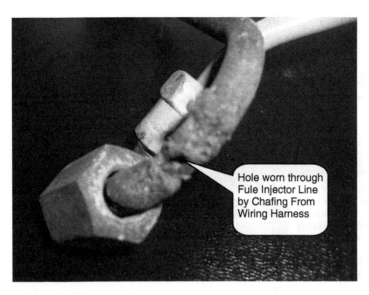

Hole worn through
Fule Injector Line
by Chafing From
Wiring Harness

Figure 5.10 Diesel injector line hole caused by chafing.

determine the cause of the fire. Once the failed injector line was identified, an immediate inspection was ordered on all buses (approximately 60 units), and it was discovered that about 10% of them had visible wear at the same location on the same fuel injector line (Figure 5.10). This very minor difference in assembly resulted in the loss of a bus, but very fortunately no injuries. Had the bus had a full load of passengers, the result could have been catastrophic. Because of this incident and the investigation into its cause, a recall was initiated and all units of that make and model were inspected and remedied (Figure 5.11).

While other failures due to unintended variations in the manufacturing process are numerous, this has been the most dramatic and significant from a potential impact on human life and injury, costliness, and difficulty in determining the root cause.

5.2.13 Factory Inspections

Factory inspections are another tool that can be used to enhance maintainability and reliability of low-volume manufactured products. Factory inspections are also a valuable tool for understanding the assembly of the unit for maintenance staff who may have to disassemble the equipment for repairs. And maintenance staff may have meaningful input into accessibility, lockout, labeling, or other insights to improve the unit's maintainability. Inspections can be done at various stages of manufacture, or even for different components. Dynamometer testing of engines, transmissions, pumps, generators, or other

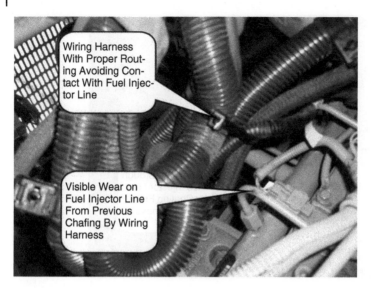

Figure 5.11 Wiring harnesses rerouted to clear injector line.

major components or systems may be appropriate for critical acquisitions. For large orders, it may be advisable to have an engineer in residence at the vendor's plant to maintain continuous overview of the assembly process.

5.2.14   Prototype Functional or Performance Testing

Consensus standards, or contractual requirements, may require first-unit or prototype testing to demonstrate conformance with the requirements. It is prudent to have an engineer or similarly qualified person to witness this functional or performance testing. If the testing is to be witnessed and attested to by an outside party, it is recommended that the services of a professional engineer be retained, and that written documentation as to the successful testing be provided and kept in the unit's file. This ensures the integrity of the testing and provides assurance to the end user that a properly designed and constructed unit is being provided. It is also valuable evidence that due diligence was employed should there be litigation at some future date. Figure 5.12 shows performance testing of a heavy-duty wrecker demonstrating its ability to make a "K" turn the narrow confines in a tunnel.

5.2.15   Acceptance Testing

Vendors do not always do a thorough final inspection of products before delivery to the customer; and even when they do, infant failures and defects during the early stages of a product's life cycle are not uncommon. A thorough

Figure 5.12 This author witnessing prototype wrecker turn around functional testing.

acceptance inspection, including checking all service fluid levels, and appropriate performance testing as soon as practical after delivery and before placing units in service is prudent. And, it can also minimize out-of-service instances that would be experienced by the operational unit had the unit been placed in service without this quality assurance. (Surprisingly the author has personally experienced production line units delivered with errors in such fundamental systems as the air brakes because of a parts mix-up during assembly.) Figure 5.13 illustrates acceptance testing of an airport snow blower. As the delivery was not during snow, load testing and noise-level measurements were performed by immersing the snow blower's head in water.

5.2.16 "Lead the Fleet" Utilization

A "Lead the Fleet" program is a tactic employed primarily by the military, in aviation, and by some civilian fleet operators, to help identify failure modes and life expectancy. This tactic maximizes the utilization of a statistically significant sample size of new fleet units to help identify failure modes and reliability issues. A select few units are doing real-world reliability testing for the other units in the production run. Although conceptually simple, to be effective such a program requires dedicated efforts on the part of all players. It requires identifying and using the "Lead the Fleet" units in the most severe operations, minimizing out-of-service (fast maintenance turnaround for both scheduled and unscheduled maintenance) for these units, and excellent data collection and analysis for the program to produce meaningful results.

Figure 5.13 This author supervising snow blower in-cab sound level acceptance testing.

5.2.17 Reserves

One of the most vexing issues for any operator of equipment, and one that is directly associated with reliability and life cycle, is what is the appropriate number of vehicles to accomplish the mission at the lowest cost. It is a problem that lacks a simple quantifiable solution. It is essentially a measure of cost and risk. How can you accurately quantify the loss due to one or more units being out of service and the probability of such an occurrence? For now, there is no answer to this question. This is one more topic that could be better answered through a more robust real-world testing process. The factors needing consideration in determining the number of units required include:

- How many units are needed to be in service to meet operational needs?
- Does this consider cyclic demands—peaks and valleys?
- How much down time is needed for scheduled maintenance?
- Can scheduled maintenance be performed during nonpeak times?
- All operations involve nonscheduled maintenance. What is a prudent estimate of nonscheduled maintenance for your organization, operation, and/or equipment? How much reserve is needed to accommodate nonscheduled maintenance?
- Repairs due to accidents, damage, or abuse for specialized equipment is lengthy and expensive. The question is what is a prudent estimate of nonscheduled maintenance for your organization, operation, and/or equipment? How much reserve is needed to accommodate accidents, damage, or abuse?

- Is there an ability to rent, lease, borrow, or some other mechanism to obtain a temporary replacement for an out-of-service unit? If available, are there operator training/compatibility issues in obtaining such a replacement?
- Will having additional units available allow for an increase in service life due to reduced duty cycle or wear and tear on operational units?
- What are the potential costs, to your organization and its operational capability should one or more units be out of service?
- What is the probability of such an occurrence?
- What is your organization's tolerance for risk?
- Who will sign off on the risk/reward or cost/benefit analysis for this decision?

5.2.18 Problem Log

An important, but often overlooked, step is documenting problems and solutions. As this is long-life equipment, it will be many years between replacement cycles. The personnel who will do the next replacement may be different due to staff changes, retirements, reorganizations, or other causes. And, the passage of time causes us to forget. A log of significant problems and solutions is a step to avoid repeating problems in the next equipment cycle. During the next replacement cycle, a careful review of the issues with the previous generation of the equipment is a powerful tool that will help to assure problems do not recur in the next generation. Unfortunately, this step is frequently overlooked or forgotten. Too often engineers are better at solving problems than in documenting them, so the narrative of what went wrong and how it was remedied is never documented and so is lost or forgotten.

5.2.19 Self-Help

Reliability and maintainability of any industrial product entails not only the manufacturer, but to a large degree it also involves the user. Training, cleaning and maintenance practices, operational practices, data collection and analysis, and planning can significantly improve or diminish the reliability and maintainability issues for this type of equipment. The following list contains some things to consider to improve the uptime of products and equipment:

- Include management, operator, and maintainer training when purchasing new units.
- Obtain new training when staff changes, and refresher training for seasonal equipment.
- If possible, schedule maintenance for off-season or off-hours to minimize impact on operations.
- Have the manufacturer provide maintenance schedule recommendations and recommended parts (items and quantity) to be inventoried. And adjust items and quantities as appropriate for your uptime needs and operational conditions.

- Combine required parts to perform scheduled maintenance tasks into kits containing all the necessary parts.
- Label fluid service locations with the proper fluids for servicing and the appropriate quantities of fluid for servicing (including fuel tank and diesel exhaust fluid tank).
- Rotate units to equalize use (unless using the previously described Lead the Fleet strategy).
- If practical, assign units to specific operators and maintenance chiefs, including putting their names on the units. Pride is a powerful motivator in keeping a unit in prime condition, and it is an inexpensive way to recognize staff.
- Collect use, repair, and parts data and use the data to develop meaningful information to improve maintainability and reliability (in-service time).
- When the units approach end of life, begin replacement activities early. End-of-life maintenance costs are very high, and timely replacement is the best way to avoid the increased out of service, repair costs, and reliability issues that result from late replacement.
- Be creative! From first-hand knowledge of the operational needs and requirements you may be able to develop other factors that can maximize unit maintainability and reliability.

References

1 National Fire Protection Association. (2012). *NFPA 414, Standard for Aircraft Rescue and Fire-Fighting Vehicles*, 2012 edition. NFPA, Quincy, MA.
2 American National Standards Institute. (2009). *ANSI/SIA A92.2 Vehicle Mounted Elevating and Rotating Aerial Devices*. ANSI, New York, NY.
3 SAE International. (2013). *SAE ARP5539, Rotary Plow with Carrier Vehicle*. SAE International, Warrendale, PA.

Exercises

5.1 List some products that are generally not subjected to laboratory or test track reliability and maintainability testing.

5.2 Identify at least three characteristics of such equipment.

5.3 Provide three examples of products that entail a high risk to an organization should the equipment fail or be unavailable.

5.4 List five strategies or tactics that can help assure the maintainability/reliability of low- volume, custom, and special-purpose vehicles and equipment.

5.5 Why are consensus standards tests not always true indicators of a product's reliability/maintainability?

5.6 Why should extended warranty and definitions of defects be included in the purchase of low-volume, custom, and special-purpose vehicles and equipment?

5.7 What are the differences between pre-award and preconstruction meetings?

5.8 Discuss several different types of testing that might be appropriate in the procurement of low-volume, custom, and special-purpose vehicles and equipment.

5.9 When should acceptance testing be performed, and why is it important?

5.10 Explain "Lead the Fleet" as a tactic.

5.11 Discuss some of the criteria that need to be considered in determining the appropriate number of units to meet operational needs.

5.12 What is a problem log, and why is it important?

5.13 Discuss several tactics that can be employed to help improve maintainability and reliability of low-volume, custom, and special-purpose vehicles and equipment.

6

Exemplary Models of Programs and Illustrations for Professional Learning in Reliability Prediction and Accelerated Reliability Testing

Lev M. Klyatis

For use in engineering education and training courses, seminars, lectures, tutorials, workshops, and so on.

6.1   Examples of the Program

6.1.1   Example 1. Several Days' Course: "Successful Prediction of Product Reliability and Necessary Testing"

Will provide the foundation for designing and using this problem solution. Course participants will learn:

First and second days: "Methodological Aspects of Reliability Prediction"

- Reliability understanding and measuring
- Impact of reliability on profits.
- How reliability influence on life-cycle cost.
- Reduction in life-cycle time and cost.
- Design FMEA.
- Misapplications of FMEA.
- Fault tree analysis (FTA).
- Conducting design reviews.
- Reliability as one from interacted component of product performance.
- Analysis of current situation with reliability of the products and the basic causes of this situation.
- Why reliability prediction is often inaccurate
- Analysis of current situation with using different methodological approaches of reliability prediction—their positive and negative aspects.
- Terms and definitions of reliability prediction.
- New advanced methodological approach for successful reliability prediction.
- The strategy of successful prediction for industry.

Reliability Prediction and Testing Textbook, First Edition. Lev M. Klyatis and Edward L. Anderson.
© 2018 John Wiley & Sons, Inc. Published 2018 by John Wiley & Sons, Inc.

- Preparation of recommendations for design improvement.
- Growth of reliability through ART.
- Controlling warranty costs.
- Implementation of successful reliability prediction.
- List of benefits, requirements, and objectives of methodological aspects of successful reliability prediction

Following days: "Accelerated Reliability and Durability Testing as Source for Obtaining Initial Information for Successful Reliability Prediction"

- Choose accelerated testing method for a given application.
- Adjust accelerated testing programs for business situations.
- Description of how product development cycles can be reduced in time.
- Analysis of current situation in methodology for accelerated field and laboratory testing.
- Analysis of current situation in equipment market for accelerated testing.
- Why the above methodology and equipment cannot offer accurate information for improving reliability prediction.
- Terms and definitions of reliability and durability testing.
- Explain the steps and methodology used to conduct of ART/ADT.
- How and why accelerated reliability and durability testing can help to solve the problem improved prediction of product reliability.
- How to utilize existing test equipment.
- Demonstrate the application of a variety of tools utilized for ART/ADT.
- How to prepare requirements to design equipment for ART/ADT.
- Implementation solution of reliability testing.
- Economic benefits from improved reliability usage.

6.1.2 Example 2. One-Day Course "Methodology of Reliability Prediction"

Will provide the methodological foundation for designing and usage reliability prediction program to achieve high reliability during given time. Seminar participants will learn:

- Why current situation with reliability prediction is often unsuccessful for practice.
- Why this situation has continued for many years (with analysis of publications).
- Terms and definitions of reliability prediction.
- How reliability prediction methodology can be improved.
- Basic components of improved reliability prediction.
- Examples with using improved reliability prediction methodology.

6.1.3 Example 3. One–Two Days' Course (or tutorial) "Accelerated Reliability and Durability Testing Technology as Source of Obtaining Information for Successful Reliability Prediction"

- Current situation with reliability and durability testing.
- Accelerated aging—positive and negative aspects.
- Highly accelerated life testing (HALT) and highly accelerated stress screening (HASS). Their advantages and disadvantages.
- The role of real world simulation in level of testing.
- The basis of accurate real world simulation.
- Key terms and definition of accelerated reliability and durability testing (ART/ADT).
- Accelerated reliability and durability testing (ART/ADT) as a key factor for improvement reliability prediction.
- Basic specifics and concepts of ART/ADT.
- Equipment for ART/ADT.
- Benefits on ART/ADT.
- Reliability testing standardization.

6.1.4 Example 4. One–Two Days' Seminar "Foundation for Designing Successful Accelerated Testing"

Will provide the foundation for designing a successful accelerated testing (AT) program so as to achieve high reliability in existing products and future designs. Seminar participants will learn:

- Why current technique and equipment for AT give no more than 20–30% of possible benefits.
- Why the simulation of real-life influences is usually not accurate for a high level of correlation between AT results and field results.
- How one can obtain maximum correlation between AT results and field results.
- How engineers and managers can find and eliminate causes for failures and degradation of product quickly and at lower costs.
- Why the quality of published literature often is not enough to turn practical engineers and managers to the way with minimum benefits.

Course participants will also learn to apply AT to:

- Shorten product time to the market.
- Reduce design and product development cycle time, warranty costs, and minimize customer returns.

AT is applicable to mechanical, electro-mechanical, electronic, hydraulic and other devices used in automotive, railroad, aerospace, marine, and other areas of industry.

Benefits of Attending

By completing this seminar, professionals will know:

- How to implement advanced and less expensive techniques to increase the product warranty period.
- Participants will be informed what advanced literature they could use for improvement of their knowledge in accelerated testing.

Seminar Basic Content

- Failure modes & effect analysis (FMEA) basics (introduction & overview, definition, general discussion, performing a design FMEA & Process FMEA, criticality/risk analysis).
- Fault tree analysis as a supportive tool.
- Introduction to successful AT.
- The strategy for creating successful AT.
- Physical simulation of real-life input influences on the product.
- Technology of step-by step AT.
- Conditions for accelerated multiple environmental testing.
- Accelerated corrosion testing with accurate simulation of field conditions.
- Vibration testing with accurate simulation of field conditions.

Who Should Attend?

Corporate executives, test engineers and managers, design engineers and technicians, reliability engineers and managers, supervisors, quality assurance managers, quality control engineers and managers, manufacturing engineers, and others.

6.2 Illustrations for these and Other Programs in Reliability Prediction and Testing

The following illustrations for the learning courses have to be prepared in slides (PowerPoint or similar software).

6.2.1 Examples: Text for the Slides

- It is a known fact that in 2009–2010 Toyota's global recall jumped to 9 million cars and trucks, later recalls could average around 3–5 million annually.
- A similar situation could happen with other automakers, as well as with companies in other areas of industry.

Figure 6.1 Introduction.

- The basic reason is the **inaccurate prediction** of the product's reliability during design and manufacturing.
- Prediction is inaccurate, **because accelerated reliability and durability testing, as source of initial information for this prediction, is not properly used.**
- Below is described how one can solve this problem.

Figure 6.1 (*Continued*)

Commonly, the situation in reliability is reflecting in recalls, because: there is official information about recalls from Government and other organizations and companies.
- In automotive industry current situation can be expressed as "**Auto Recalls Accelerate.**"
- The US Federal Government (NHTSA) said **In 2011 automakers recalled more US vehicles last year than in any of the last six years.*" and
- "**Recalls affected** 20.3 million vehicles, the **highest number since 2004**" (2009, 15.2 million).

Figure 6.2 Current situation with product reliability.

In the field reliability of equipment for the military area are several times lower than was predicted after testing during design and manufacturing.
*SAE International G-11 RMSL Division Spring 2004 Meeting.

Figure 6.3 Brigadier General Carl Schenk described.*

- Use examples with newest information from NHTSA (National Highway Safety Traffic Administration, DoT or other publications).

Figure 6.4 There are many other examples with recalls.

1. Honda – ?.? million vehicles.
2. Toyota – ?.? million.
3. Ford – ?.? million.
4. Other – ?? million.
**a. Source…
b. The companies could be changed.

Figure 6.5 Highest recalls during last year.*

- HIGHER COST
- LOWER SAFETY
- LOST TIME
than predicted during design & manufacturing.

Figure 6.6 Not only recalls, but.

YEARLY FATALITIES AND FATALITY RATE (NHTSA): (deaths: 2010, 33,186; 2011, 32,310), (other years).

Figure 6.7 One from final result of inaccurate prediction is.

Efforts of the industrial companies for reducing recalls and life-cycle cost, increasing quality, reliability, safety, durability, maintainability, and others.
- What are the causes?
- Present approaches do not offer the possibility for successful solution to the above problems.

Figure 6.8 It is real fact that.

Using:
- Engineering technology (design, technology, quality control, etc.) analysis
- Physical analysis
- Chemical analysis
- Statistical analysis

Figure 6.9 Ways for finding the causes for complaints and recalls.

Phillip Coyle, the former director of the Operational Test and Evaluation Office (Pentagon) said that:
- **if during the design and manufacturing** complicated apparatus such as satellite, <u>one tries to save a few pennies in testing,</u>
- **the end results may be a huge loss of thousands of dollars due to faulty products which have to be replaced, because of this mistake.**

Figure 6.10 Example: results of saving expenses for testing during design and manufacturing.

The reliability function of distribution after ART is $F_A(x)$, in the field is $F_0(x)$.
The measure of their difference is
$\Delta[F_A(x), F_0(x)] = F_A(x) - F_0(x)$
for $\Delta[F_A(x), F_0(x)]$ given limitation is Δ_A.
If $\Delta[F_A(x) - F_0(x)] \leq \Delta_A$
it is possible to determine the reliability of ART results.
But if $\Delta[F_A(x) - F_0(x)] > \Delta_A$
it is not recommended.

Figure 6.11 Statistical criteria for comparison the reliability in results of accelerated reliability testing (ART) and field testing.

If one doesn't know the functions $F_A(x)$ and $F_0(x)$ (often in practice) one can construct the graphs of the experiment data $F_A(x)$ and $F_0(x)$ and determine

$$D_{M,N} = [F_{AE}(x) - F_{OE}(x)]$$

where

F_{OE} and F_{AE}

are empirical distributions of the reliability function as result of the machinery testing under operating conditions and by ART/ADT.

Figure 6.12 Statistical criteria ... (continuation).

One needs to calculate the accumulated parameter's function and the values of the confidence coefficient found in the equations:

$$Y(x) = \sum_{n}^{m=k} C_n^m p^m (1-p)^{n-m}$$

$$Y(x) = \sum_{k}^{m=k} C_n^m p^m (1-p)^{n-m}$$

And evaluate the curves that are limited to the upper and lower confidence areas, where $C_n^m p^m (1-p)^{n-m}$ is the probability that based on an event will be in n independent experiments m times. The values of Y are found in the tables of books on the theory of probability if the confidence coefficient is

$$Y = 0.95 \text{ or } Y = 0.99.$$

Figure 6.13 Comparison parameter's function with predetermined accuracy and confidence area.

More stress means greater acceleration and lower correlation of accelerated testing results with field results

Figure 6.14 Axiom of stress testing.

Figure 6.15

Basic Concepts of ART/ADT

1. <u>Accurate simulation of the whole complex</u> of the field situation, using given criteria.
2. Simulation of field input influences includes:

   A. Simulation for testing 24 hours a day every day, but not including:
   - idle time (breaks, etc.). (What that means? Example with car);
   - time with minimum loading which does not cause failures (using "careful stress" method, [see the details in author's books]);

   B. The simulation of environmental influences (temperature, pollution, radiation, etc.) is providing with "special maximum stress" method.

   For simple product (for example, mixer) this method will be different.

Basic Concepts of ART/ADT (cont. 1)

Figure 6.16

c. Above is a description of a general approach to the method of accurate simulation. If we need to use this for the specific product, for example, cars, we will create a specific method of accurate simulation input influences, for the mixer we need another specific method, etc.

D. A similar situation relates to accurate simulation interconnected human factors and safety.

E. As a result of the above approach use, the acceleration factor is usually from 10 to 100 and more (depending on the type of the product).

3. Simulation simultaneously and in combination for each group of field influences (multi-environmental, mechanical, electrical, and other).

BASIC CONCEPTS of ART/ADT (cont. 2)

Figure 6.17

4. Systems must be treated as interconnected, using

Systems of Systems approach which consists of accurate

simulation and interconnection of each of the field input

influences, safety, and human factors.

5. Accurate simulation of the simultaneous combination of each

type of input influence, using given criteria.

For example: Pollution = chemical + mechanical (dust, sand).

6. Use of the physics-of-degradation mechanism as a basic

criterion for accurate simulation of the field influences.

BASIC CONCEPTS of ART/ADT (cont. 3)

Figure 6.18

7. Consider components (of test subject) interaction within the system.

8. Reproduction of the complete range of field schedules and maintenance (repair).

9. Laboratory testing in combination with special field testing as basic components of ART.

10. Corrections of the simulation system after an analysis of the field degradation and failures and compare this with the degradation and failures during ART.

Figure 6.19

Benefits of ART/ADT
Life Cycle Costing (LCC), %

Results of implementation described development of ART/ADT,%:
- increasing the cost of design phase – 4-8 %*;
  (not including the benefits from recalls and maintenance reducing)
- increasing the cost of manufacturing phase – 0-1 %*;
  (not including the benefits from recalls and maintenance reducing)
- decreasing the cost of usage phase – 52-83.

As a result, total LCC decreases minimum by 33-47 %

%*The cost of design and manufacture phases will decrease if the industrial company repeatedly uses ART/ADT equipment for the future models of the product, as well as manufacturing phase.

Figure 6.20

Examples for ART/ADT Suitable Testing
Equipment (Now in the world market)

- **WEISS TECHNIK (Germany):**

1. **Climate Test Chamber with Road Simulator and Solar Radiation:**
   This system simulates and combines vibration with environmental conditions such as humidity, heat, cold, and solar radiation.

2. **Combined Corrosion Testing System:** The system comprises two operating reliability test beds and enables the combination of the test parameters – temperature, humidity, corrosion with NaCl; CaCl, and MgCl solutions while simultaneously being subject to mechanical load in three axes.

Figure 6.21

Examples of ART/ADT Suitable Testing Equipment

- Seoul Industry Engineering Co., Ltd. R&D Center

Bus Climatic Wind Tunnel simulates:
simultaneously combination of temperature, humidity, solar light, chassis dynamometer, vibration, wind speed and flow.

- State Enterprise TESTMASH (Moscow, Russia):
Reliability/Durability Test Chamber simulates: temperature, humidity, radiation, pollution, vibration, dynamometer, input voltage.

Example. ART/ADT ACCELERATED THE PRODUCT DEVELOPMENT & FINDING of FAILURES CAUSES

- The designers did not eliminate the problems with the harvester's reliability and durability over several years of field testing and data collection.
- A special complex for field simulation was developed by the Dr. L. Klyatis laboratory.
- After 6 months of ART:
- Two of the harvester's specimens were subjected to prediction for the equivalent 11 years.
- Three variations of one unit and two variations of another unit were tested.
- reliability and durability was increased over 2 times.

Figure 6.22

Expenses During the Life Cycle of Machinery

Figure 6.23

The experience shows:

- If industrial companies during the design process will

    use ART/ADT),
    this will increase reliability and

    decrease recalls and complaints.

- As a result, dramatically decreasing expenses during total design, manufacturing, especially during usage.

Figure 6.24

IMPLEMENTATION of ART/ADT

- The beginning of implementation of ART/ADT technology requires a longer time for the first model of the product (test subject).

- The implementation time will be reduced for the following models of the product. The reducing process will be continuous during ART/ADT usage time.

- If management has concerns that the above technology of ART/ADT requires too great an investment, the implementation can be accomplished through a step-by-step implementation over time:

-- first, build separate room with the basic communications (for water and other liquid transportation, gas, etc.) and a drainage system;

-- then, inside of this room implement the test equipment for one type of testing, for example, vibration.

-- later when funds became available, add a second type of equipment to the test facility implementing temperature variation and its interaction with vibration levels;

-- continue adding other influencing variables.

Figure 6.25

USE THE RESULTS OF ART/ADT

- One can use directly for **evaluation** of reliability and durability for the conditions of testing, i.e. for the laboratory conditions, but **not** for real field conditions.

- If one wants to know the reliability and durability in **real world** after these testing, one needs **to predict** this, using **prediction methods**.

Figure 6.26

ACCURATE RELIABILITY PREDICTION

- Using the **sufficient initial information,** obtained after ART/ADT and current **advanced methodologies of prediction**, one can provide a successful reliability, prediction, as well as their effective development & improvement.

- There is ART/ADT methodology, developed by Dr. L. Klyatis.

- One can see this detailed methodology in his publications.

THE PREDICTION WILL BE ACCURATE:

Figure 6.27

• If the initial information is ACCURATE.

• But separate types of influences simulation and used stress testing, with simulation only part of the field situation, do not offer the possibilities to obtain this accurate information.

FOR AN SUCCESSFUL PREDICTION

Figure 6.28

• One needs to use accelerated reliability/ durability testing as a **key factor** for obtaining corresponding initial information.
• One has to know <u>how to do this</u>, because this type of testing is very seldom provided.
• Now during design and manufacturing one uses HALT, Accelerated Aging, HASS, software simulation approach, etc., because they are simpler and cheaper for testing providing (**only**).
• Will be prediction successful or life cycle cost cheaper using the above? No!

6.2.2 Examples of Figures

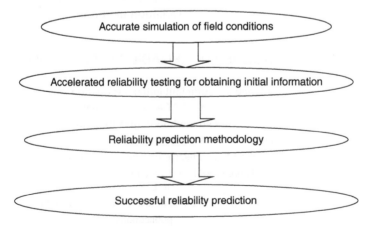

Figure 6.29 Four basic steps for successful reliability prediction.

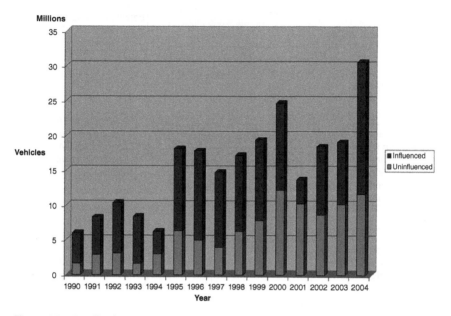

Figure 6.30 Recalls of automobiles from 1990 to 2004 (millions) in the American market.

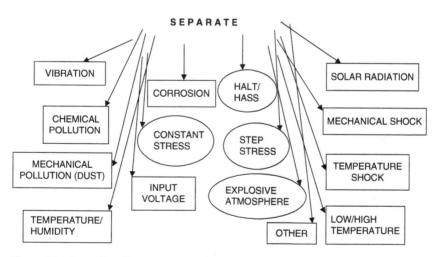

Figure 6.31 Examples of separate types of practical simulation and stress testing during design and manufacturing. This is low effective way.

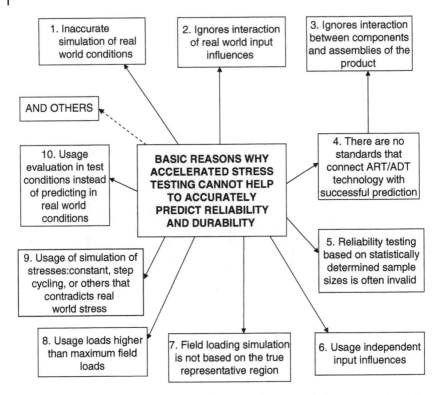

Figure 6.32 Basic reasons why accelerated stress testing cannot help to accurately predict reliability and durability.

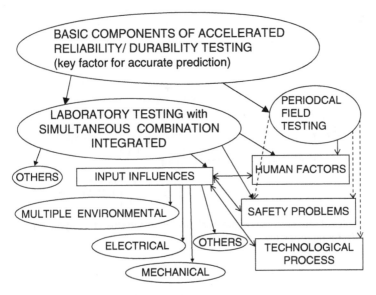

Figure 6.33 Basic components of ART/ADT.

EXAMPLE OF PERIODICAL FIELD
TESTING

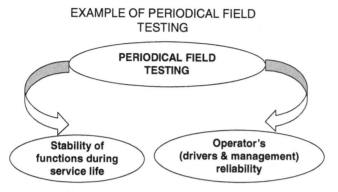

Figure 6.34 Example of periodic field testing.

EXAMPLE OF INTEGRATED (SIMULTANEOUS
COMBINATION) OF THE <u>REAL WORLD</u> INPUT
INFLUENCES ON THE PRODUCT

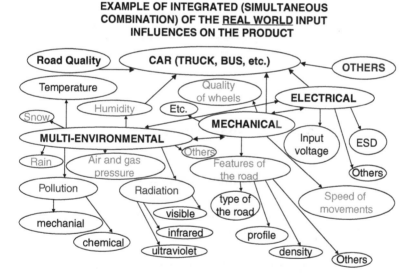

Figure 6.35 Example of interacted (simultaneous combination) of the real-world input influences on the product.

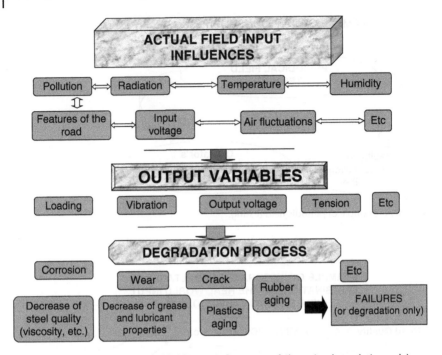

Figure 6.36 The way from actual field input influences to failures (or degradation only).

Example. **The Types and Parameters of the Degradation Mechanisms**

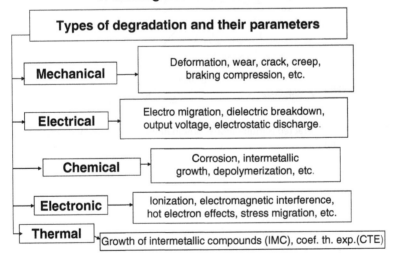

Figure 6.37 Example: The types and parameters of the degradation mechanisms.

THE WAY TO RELIABILITY/ DURABILITY TESTING

• Steps of ART/ADT development:

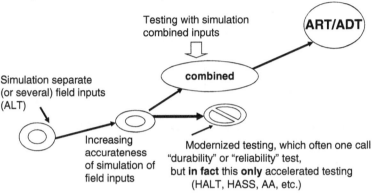

Figure 6.38 The way to reliability/durability testing.

Contents of ART/ADT Technology

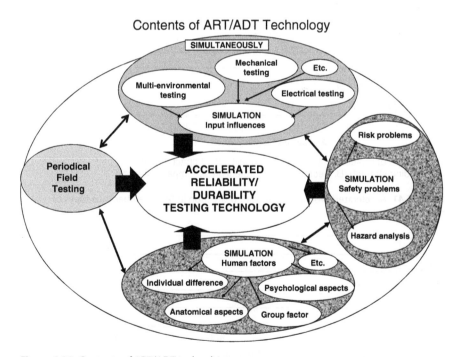

Figure 6.39 Contents of ART/ADT technology.

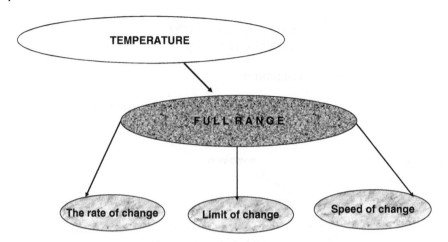

Figure 6.40 Scheme of study the temperature as an example of accurate simulation input influences.

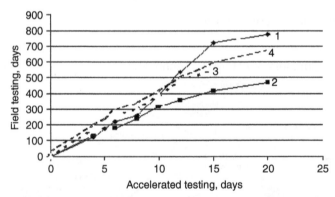

A. First type of protection: 1 – protection quality; 2 – impact strength; 3 – bending strength. B. Second type of protection: 4 – impact strength.

Figure 6.41 Accelerated destruction of paint protection in test chamber (two types of paint).

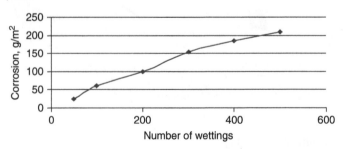

Figure 6.42 Dependence of steel corrosion values on the number of wettings in test chamber.

Figure 6.43 Vibration in test certification process in aircraft. The 190 Aircraft and vibration equipment.

Figure 6.44 The system (test subject) as complex of interconnected components (units and details).

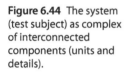

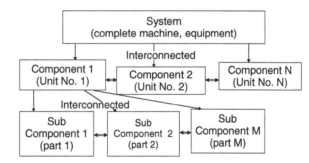

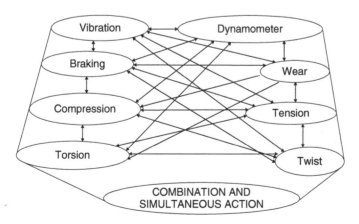

Figure 6.45 Different types of mechanical testing.

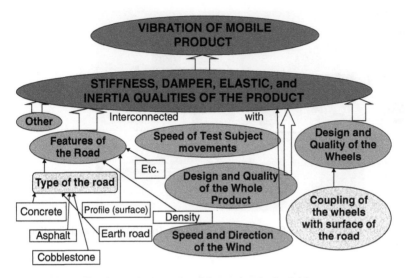

Figure 6.46 Technology vibration of mobile product in the field.

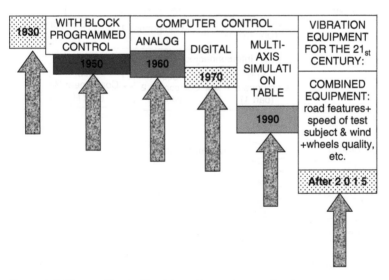

Figure 6.47 Stages of basic vibration testing equipment development.

Figure 6.48 Climate test chamber with four-wheel-drive dynamometer with sunlight simulation (Weiss Technik).

Figure 6.49 Cold-head climate test chamber with road simulator and sunlight simulator (Weiss Technik).

Figure 6.50 Combined testing system: vibration, climate, and corrosion (Weiss Technik).

Figure 6.51 Combined test chamber for electronic devices. Simulates vibration, temperature, input voltage, and humidity.

Figure 6.52 Bus climatic wind tunnel.

Bus Climatic Wind Tunnel
Seoul Industry Engineering Co., Ltd.
R&D Center

- ## SPECIFICATIONS:

- Temperature control range   −40°C ÷ 60°C
- Humidity control range   10 ~ 90% RH
- Solar light control range 0 ~ 1400 w/m^3 (0 ~ 1200 cal/m^2·hr)
- Chassis dynamometer power control range   0 ~ 373 kw
- Wind speed control range   0 ~ 100 mph
- Make up air   Provided at − 40°C
- Wind flow uniformity   3.0% at 80% nozzle area
- Available for vibration
- Boiler heating & steam injection system

Figure 6.53 Bus climatic wind tunnel: specifications.

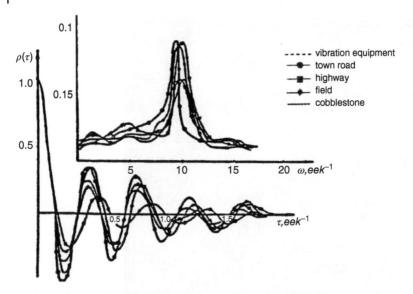

Figure 6.54 Normalized correlation and power spectrum of frame tension data for the car's trailer in different field conditions and in the chamber.

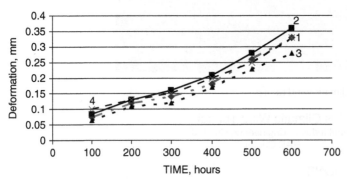

1– mean deformation in the field; 2 – upper confidence limit in the field;
3 – confidence limit in the field; 4 – mean deformation during ART.

Figure 6.55 Deformation of metallic sample during the time in the field and during ART/ADT.

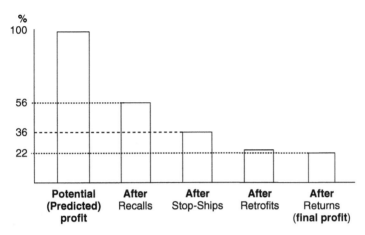

Figure 6.56 Effect of poor reliability on profit.

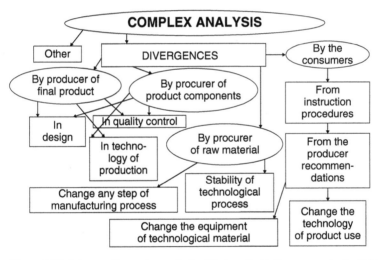

Figure 6.57 Scheme of complex analysis of factors that influence product reliability/quality.

Index

Reliability Prediction and Testing Textbook, First Edition. Lev M. Klyatis and Edward L. Anderson.
© 2018 John Wiley & Sons, Inc. Published 2018 by John Wiley & Sons, Inc.